Proceedings in Life Sciences

Approaches to New Leads for Insecticides

Edited by
H. C. von Keyserlingk, A. Jäger,
and Ch. von Szczepanski

With 68 Figures

Springer-Verlag
Berlin Heidelberg New York Tokyo

Dr. HARALD C. VON KEYSERLINGK

Dr. ALBERT JÄGER†

Dr. CHRISTOPH VON SZCZEPANSKI

Schering AG
Agricultural Research
P.O. Box 650311
1000 Berlin 65, FRG

ISBN-13:978-3-642-70823-7 e-ISBN-13:978-3-642-70821-3
DOI: 10.1007/978-3-642-70821-3

Library of Congress Cataloging-in-Publication Data. Main entry under title: Approaches to new leads for insecticides. (Proceedings in life sciences). Includes index. Contents: The challenge of finding new insecticides for a mature market/U. Eder and H. C. von Keyserlingk – The biochemical mode of action of insecticides/A. C. Baillie – Can mode of action studies guide chemical synthesis?/T. A. Miller – [etc.] 1. Insecticides–Congresses. 2. Insecticides–Physiological effect–Congresses. I. Keyserlingk, Harald C. von, 1948– . II. Jäger, Albert, 1925–1985. III. Szczepanski, Christoph von, 1937– . IV. Series. SB951.5.A66 1985 632'.951 85-27628

2131/3130-543210

Preface

Man as a biological species and as a member of the community of organisms that has evolved over the billions of years on this planet has to face competition for his resources in the same way as any other species. In the course of evolution organisms have developed a fascinating variety of weapons, techniques and strategies in this competitive scramble for food and other essential resources and in self-defense against parasitism and predation. Among these, the biosynthesis of toxic molecules and their appropriate delivery in interspecific encounters has proved to be one of the most successful natural weapons and has been demonstrated in every branch of the evolutionary tree of all extant species.

It is interesting to note that in this process of generating toxic molecules both natural selection by evolutionary processes and screening techniques by chemical industry have produced compounds with very similar target sites. The most potent of these compounds interfere with the normal signalling of nerve cells. Saxitoxin, tetrodotoxin, batrachotoxin, grayanotoxin, veratridine, aconitin, sea anemone venom and scorpion toxin are natural poisons which interfere with the same membrane bound protein as many potent synthetic insecticides such as DDT and the pyrethroids: the voltage-gated Na^+ channel in the nerve cell membrane. In several cases natural toxins have directly provided lead structures for successful chemical synthesis programmes in the insecticide industry. Eserine, nereistoxin and pyrethrum are outstanding examples.

However, today, 45 years after the discovery of the insecticidal potency of DDT we realize that the subsequent introduction of nearly 200 synthetic insecticides, has been based on a very limited number of lead structures.

In our commitment to develop highly effective, environmentally safe and commercially successful compounds for modern crop protection and public and animal health we feel that new, original lead structures are certainly needed. The challenge today is to learn from nature, to discover potential target sites and to identify the toxophores in the often bulky, complex and unstable structures of the many natural, bioactive molecules. The successful, recent development of the synthetic pyrethroids clearly demonstrates the potential of this approach in terms of scientific and economic rewards. A great variety of natural toxins and other pharmacological tools is known to interfere with many more neurochemical targets than have been exploited so far, e.g. transmitter- or voltage gated ion channels, the pathways for transmitter or neurohormone synthesis degradation or re-uptake, the secondary messenger systems regulating the concentrations of cyclic nucleotides inside excitable cells, to mention only a few.

In our attempt to identify new lead structures for insecticide synthesis we realised that we should broaden our views and discuss our ideas with some of the leading scientists in this area of insect toxicology and receptor pharmacology. In these scientific disciplines it is daily practice to study the responses of living organisms, tissues, subcellular fractions or single proteins to endogenous signals, drugs, toxins, insecticides or any other bioactive molecules. Sometimes these pharmacological probes and tools are modified with the purpose of manipulating their binding to receptor proteins: for example, to induce covalent binding.

With this broad objective in mind and in the context of the completion of new laboratory buildings for our agrochemical research, Schering AG invited leading members of the scientific community to this symposium, which took place in Berlin, on August 26 to 28, 1984. We enjoyed stimulating discussions between academia and industry on potential targets, potential leads and appropriate research techniques. We feel that the presentations during this meeting are fundamental enough to be of interest to a larger audience than we were able to accommodate. So we are particularly grateful that Springer Verlag offered to publish the contributions to our symposium in this series Proceedings in Life Sciences.

Our coeditor, Albert Jäger, head of biological research in the agrochemical division of Schering AG, died on August 20 1985 at the age of 60. This unexpected and sudden loss has been a shocking experience for all of us who had contact with him. The size of the gap that his death has created made us realise what a warm and humble personality he was, and demonstrated the deep influence that he had in shaping biological research in our agrochemical division. He joined this organisation in 1954, and from that time entomological, and later biological research in general, grew and expanded greatly under his influence.

Recently, our organisation has made great efforts to incorporate today's expertise in the fast advancing fields of physiology and biochemistry into our agrochemical research. Albert Jäger was an initiator and committed promoter of this development.

We would like to dedicate this volume to the memory of this outstanding man, who, for all his quietness, was brilliant in his scientific and managerial talents and achievements, and an invaluable guide to those of us who knew him more closely.

Berlin, October 1985

H. C. von Keyserlingk
Ch. von Szczepanski

Contents

Contents

List of Contributors

You will find the addresses at the beginning of the respective contribution

List of Abbreviations

α-BGT	α-bungarotoxin
α-BCP	α-bag cell peptide
ACh	acetylcholine
AChE	acetylcholine-esterase
AChR	acetylcholine-receptor
ADP	adenosine diphosphate
α-GPI	NAD-linked α-glycerophosphate dehydrogenase
α-GPII	α-glycerophosphate dehydrogenase
AJH	anti-juvenile hormone
AKH	adipokinetic hormone
AMP	adenosine monophosphate
ATP	adenosine triphosphate
Barb	barbiturates
BHC	hexachlorocyclohexane
Bz	benzodiazepine
CC	corpus cardiacum
CDM	chlordimeform
CDNA	complementary desoxyribonucleic acid
ChAT	choline acetyl transferase
CNS	central nervous system
CoA	coenzyme A
DCDM	desmethyl-chlordimeform
DDE	1,1-dichlor-2,2-bis(p-chlorphenyl)ethylene
DDT	dichlorodiphenyltrichlorethane
DFP	diisopropylphosphofluoridate
DG	diglyceride
DHAP	dihydroxyacetone phosphate
DLM	dorso longitudinal muscle
DMAPP	dimethylallyl pyrophosphate
DNA	desoxyribonucleic acid
Ds	slow coxal depressor motorneuron
DUM	dorsal unpaired median neuron
DUMETi	dorsal unpaired median extensor tibialis
ED_{50}	effective dose for 50% of the test population
EDTA	ethylenediaminetetraacetic acid
ELH	egg laying hormone
ELISA	enzyme linked imunosorbent assay

EM	electron microscope
EMAPP	ethylmethylallyl pyrophosphate
EMD	ethyl (E)-3-methyl-2-dodecenoate
EPSC	excitatory postsynaptic current
EPSP	excitatory postsynaptic potential
ETB	Ethyl-4-[2-(tert-butylcarbonyloxy)-butoxy]benzoate
FAB	fast atom bombardment
FAD	flavin adenine dinucleotide
FGP	mitochondrial flavoprotein
FLU	flunitrazepam
FMev	fluoromevalonate
FMN	flavin mononucleotide
Fp	succinate dehydrogenase
FPD	NADH dehydrogenase
γ-HCH	γ-hexachlorcyclohexane
GABA	γ-aminobutyric acid
GMP	guanosine monophosphate
GTP	guanosine triphosphate
HIPP	homoisopentyl pyrophosphate
HMev	homomevalonate
HPLC	high performance liquid chromatography
5-HT	5-hydroxytryptamine
IBMX	isomethylbutylxanthine
IC_{50}	concentration that produces 50% inhibition
Ins 1P	inositol 1 phosphate
IP3	inositol-1,4,5-triphosphate
Ipp	isopentenyl pyrophosphate
IUPAC	International Union of Pure and Applied Chemistry
JH	juvenile hormone
Kd	dissociation konstant
Kdr	knock down resistance
K_i	inhibition konstant
K_t	transport konstant
LD_{50}	lethal dose for 50% of the test population
M	mole
mEPSP	miniature excitatory postsynaptic potential
Mev	mevalonate
mg	milligram
M I	myoactive peptide I
M II	myoactive peptide II
mM	millimole
mRNA	messenger ribonucleic acid
msec	millisecond
mV	millivolt
$N\text{-}HF_e$	specific protein containing nonheme iron
NAD	nicotin adenine dinucleotide
NADP	nicotin adenine dinucleotide phosphate

ng	nanogram
nM	nanomole
PB	piperonyl butoxide
PCP	phencyclidine
ppm	part per million
Ptd Ins	phosphatidyl inositol
PTX	picrotoxinin
QNB	quinuclidinyl benzilate
RC	respiratory control
RPCH	red pigment concentrating hormone
SDH	succinate dehydrogenase
SDS	sodium dodecyl sulfate
SETi	slow extensor tibialis motorneuron
SR	sarcoplasmic reticulum
TBPS	t-butylbicyclophosphorothionate
TCA	tricarboxylic acid
ThG	thoracic ganglion
TPMP	triphenylmethylphosphonium
TTC	2,3,5-triphenyl tetrazolium chloride
μg	microgram
μl	microliter

The Challenge of Finding New Insecticides for a Mature Market

U. EDER and H.C. VON KEYSERLINGK [1]

1 Introduction

Contemplating our theme "Approaches to New Leads for Insecticides" and considering the current status of the highly developed insecticide industry I would like to comment on some of the incentives for this search for new leads which is demanding so much effort in intellect, manpower and capital. Are there really not enough excellent products already on the market? Where are the yet unsolved insect problems in agriculture and public health? Is it really the humanitarian desire to feed the poor of this world or is it just the scramble for a share of the market by competing companies that fuels this race for evermore active compounds?

In order to communicate my view on this subject I would like to divide the title of this contribution into the two questions:

- What are the challenges in finding new insecticides?
- How mature is the market?

2 The Challenges in Finding New Insecticides

As a research chemist working over the past 15 years on the synthesis of new molecules I confess that to me it will always be a great intellectual challenge to discover new compounds with hitherto unknown biological activities.

Sometimes, an original lead structure opens the door for the development of completely new classes of active compounds. In that case both the chemist and the biologist will experience the satisfaction and pride of the true discoverer and inventor. In this sense Paul Müller is the discoverer of DDT and, indeed, he won the Nobel price for this achievement.

Similarly, Michael Elliott is the inventor of the synthetic pyrethroids and he, too, has won several distinguished awards.

Such highlights are very rare events. Nevertheless, many effective compounds have found their way on to the market.

Probably the most common way in which chemical research is done today in the area of plant protection is the synthesis in analogy to already known lead compounds. Such a lead may be a natural toxin such as the pyrethrins or nereistoxin but more likely it will be an existing insecticide, a patented research compound of our competitors or a lead picked up in our own random screening.

1 Schering AG, Berlin/Bergkamen, Postfach 65 03 11, 1000 Berlin 65, FRG

Approaches to New Leads for Insecticides
Edited by von Keyserlingk et al.

In all theses cases, of course, we are trying to exploit niches and known weaknesses of existing compounds in terms of potency, activity spectrum or selectivity. Today, the modification of the alcohol moiety of pyrethroid esters may not seem to be an outstanding intellectual advance. However, if such a new substituent added a reasonable acaricidal activity to the excellent insecticidal properties of the pyrethroids and at the same time reduced the fish toxicity this would have to be regarded as a great achievement and success. Such successes do occur much more frequently than the discoveries of entirely new classes of insecticides.

Therefore, this approach of taking small steps has been a relatively safe way to substantial progress until now and I believe that it still has reasonable potential and that even the already existing leads still bear considerable unexplored potential. However, the eager chemist will always seek the challenge of exploring and discovering new, original chemical structures.

We should bear in mind, that the large majority of all existing insecticides were discovered by so-called random screening of substances which had not been designed in a particularly rational way. Until now the fine mesh of careful screening has filtered out interesting lead compounds as well as marketable insecticides. Most insecticides, DDT, many organophosphates and carbamates as well as the acylureas such as dimilin were discovered this way. The random screening of microbial fermentation broths has led to the discovery of the extremely potent avermectins.

However, today we are faced with the fact that the mere probability of a compound reaching the market has decreased dramatically: It is not more than one out of 10–15,000. Some authors even expect this figure to rise to about 70,000 by 1990 (Geissbühler 1984).

Therefore, it is not just intellectual ambition that makes our profession susceptible to the idea of a more rational design of bioactive molecules.

This effort to synthesize molecules by designing them to interfere with particular proteins in the target species has received increasing attention in recent years (Magee et al. 1984). This type of approach has proved to be a formidable challenge. The requirements that have to be met by a new bioactive molecule if it is going to reach the market as an insecticide are very high. It should be stable enough to exert its activity in the field yet it should be sufficiently labile to prevent environmental contamination. It should act selectively against the target species, if possible also against those populations that have developed resistance against existing insecticides. The toxicity of a compound is the consequence of complex interactions between the molecule and the contaminated organism. Target site activity and the availability at the target site which is determined by factors like penetration, storage and excretion and metabolic conversion are crucial. Therefore, several, independent factors have to be considered when attempting to design a molecule for a specific purpose.

Even though this approach is still short of convincing success we are eager to absorb and incorporate as much of the current knowledge and expertise in insect physiology, biochemistry or molecular biology into our synthetic efforts as we can.

Today, a large number and variety of very effective insecticides is available. Against this background the discovery and the development of insecticides with entirely new and different modes of action is a great challenge.

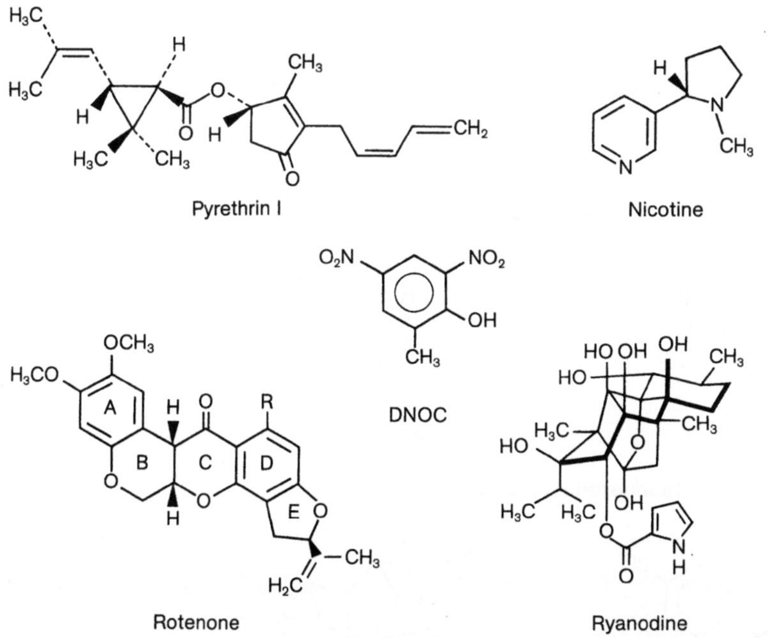

Fig. 1. The structures of some of the earlier insecticides, sometimes called the first generation insecticides

Farmers used natural insecticides such as pyrethrins, nicotines, rotenone or ryanodine, inorganic chemicals such as lead arsenates or tar oils decades or even centuries ago. The dinitrophenols which were introduced about 100 years ago were the first synthetic organic insecticides.

All these compounds together form a very divergent group with some sort of insecticidal activity. They had been the only alternative to purely cultural pest control operations. Altogether they are usually called the first generation insecticides.

Generally, the efficiency and the selectivity of these products against insects is somewhat limited, but they were the best compounds available until about 1940. It is interesting to note, however, that some insect populations had already developed resistance against some of these compounds during early decades of this century (Melander 1914, Barbers and Pratt 1951, Forgash 1984).

A dramatic increase in insecticidal potency of synthetic materials began around 1940 with Geigy's introduction of DDT. At about the same time Schrader of Bayer discovered the insecticidal properties of the organophosphates and later the first carbamate insecticides have been introduced, again by Geigy.

These developments had a profound impact on agriculture and public health: Productivity has increased substantially and diseases such as malaria were eradicated from entire geographic areas. These groups of highly potent neurotoxins dominated the insecticide market throughout the past 40 years and still do so to-

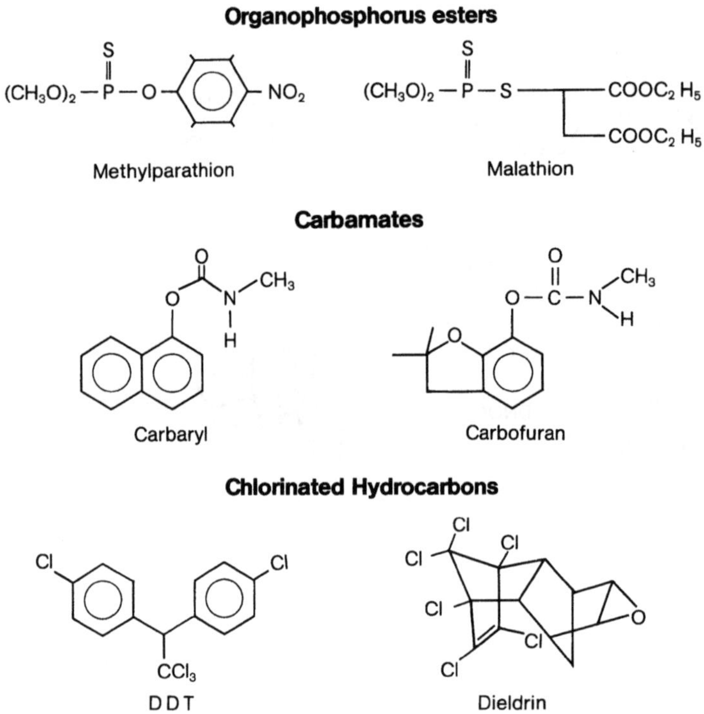

Fig. 2. The structures of some of the so-called second generation insecticides

day. About ten years ago, after the withdrawal of several persistent chlorinated hydrocarbon compounds the very potent synthetic pyrethroids were introduced.

Together these large groups of very efficient, neuroactive compounds, the chlorinated hydrocarbons, the organophosphates, the carbamates and the pyrethroids, are usually called the second generation of insecticides (Hummel 1983). Various members of these groups with very divergent insecticidal properties and environmental behaviour have been the best sellers over the past 40 years or so. In recent years the synthetic pyrethroids have conquered a substantial share of the market.

Initially, high potency and broad spectrum activity were most important. Later other aspects such as reduced toxicity against vertebrates and entomophageous insects or reduced environmental stability became desirable.

It is important to note that virtually the entire insecticide industry as it stands today has been built on these potent molecules which have in common that they modify the normal signalling of nerve cells. Here they interfere either with the proteins of the voltage-gated sodium channels in nerve cell membranes producing a prolonged sodium influx into the cells as DDT and pyrethroids do, or they interfere with synaptic process, for example, by binding to the acetylcholine-esterases and/or acetylcholine-receptor leading to an increased opening of the post-synaptic acetylcholine-gated cation channels, as organophosphates and carbamates do.

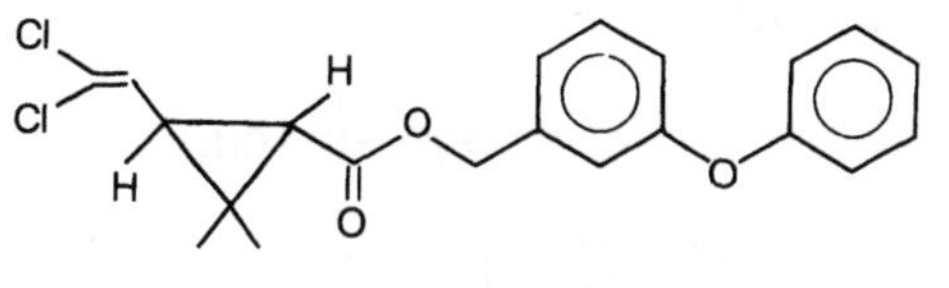

NRDC 143 (Permethrin)

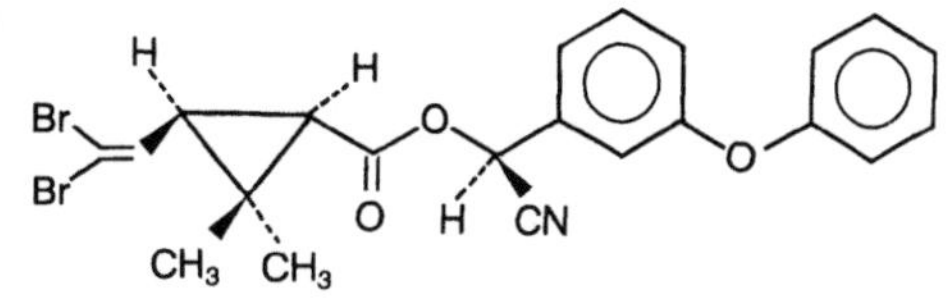

NRDC 161 (Decamethrin)

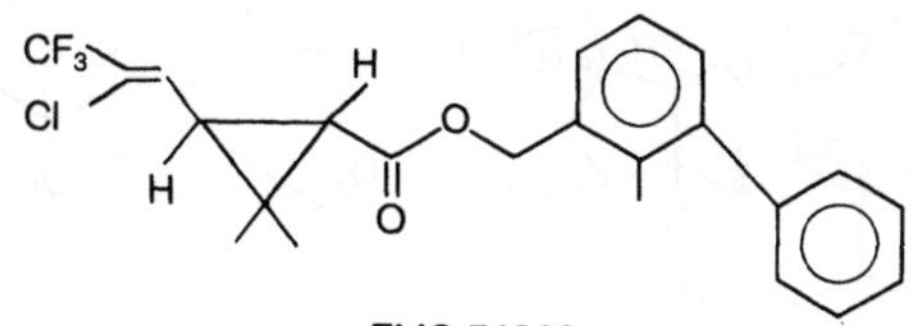

FMC 54800

Fig. 3. The structures of some synthetic pyrethroids

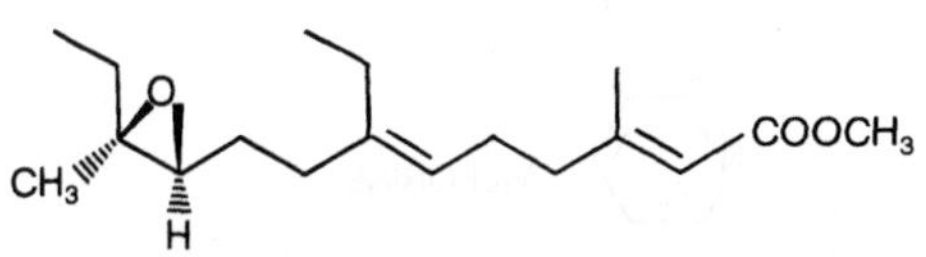

JH I

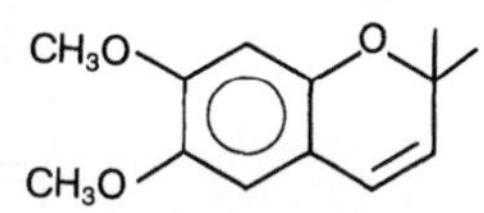

Precocene II

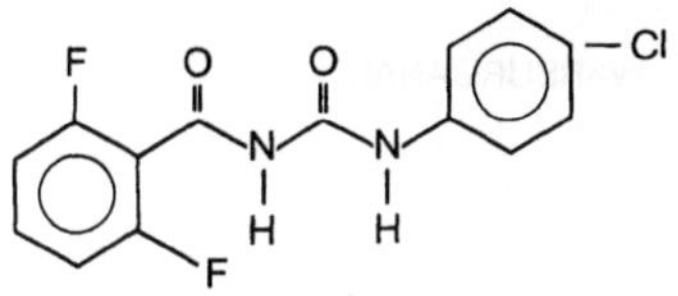

Dimilin®

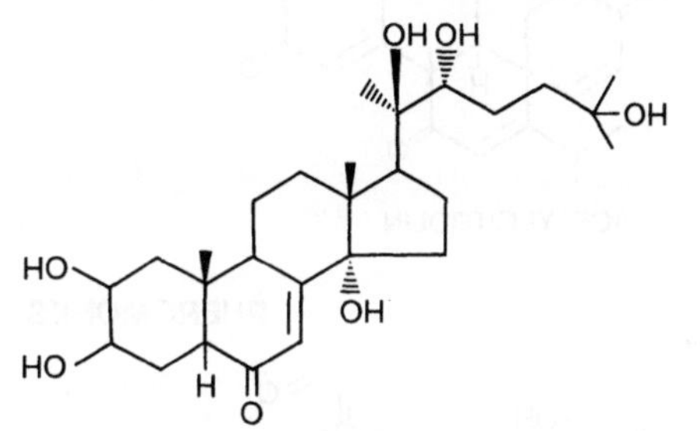

β-Ecdysone

Fig. 4. The structures of some insect hormones or compounds that interfere with the normal development of insects, sometimes termed third generation insecticides

These two basic biochemical systems are common to virtually all animals. Nevertheless, individual members of these insecticide groups can be rather selective, sparing mammals, e.g. malathion, or beneficial insects, e.g. pirimicarb. This indicates that properties such as selectivity are not necessarily determined by the target site activity of these molecules but rather by other factors such as rates of penetration and metabolism, for example.

After the achievement of potency and efficiency in the 1940's, 1950's, and 1960's, research began to focus more and more on factors such as selectivity and environmental safety. In the late 1960's a third generation of insecticides was born: the insect growth regulators.

Compounds disturbing hormone signalling or chitin synthesis such as juvenile hormone analogues, ecdysones, precocenes and acylureas are generally fairly selective against insects and environmentally rather safe. However, they act slowly,

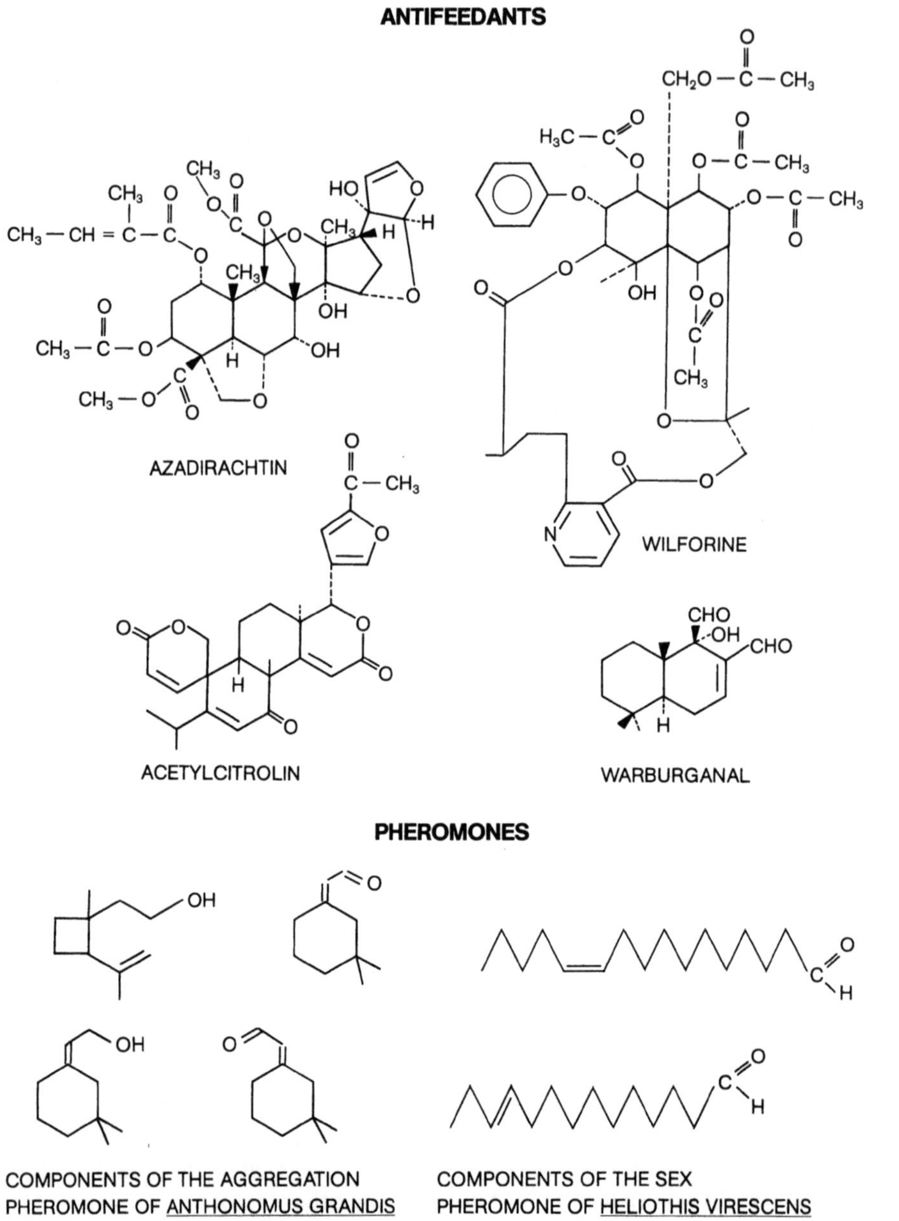

Fig. 5. The structures of some compounds with antifeedant activity and some insect pheromones, sometimes called fourth generation insecticides

often allowing the insects to consume considerable quantities of their host plants before they die. We have to realize the fact that throughout the past 15 years or so these compounds have hardly been capable of gaining a substantial share of the market – in sharp contrast to the recently introduced synthetic pyrethroids.

I believe that unless we get these compounds to act faster or at least to stop the feeding of the treated insects their position in the market will remain a small niche.

We see a comparable picture in yet another group of compounds, sometimes called the fourth generation of insecticides: the so-called behaviour-modifying chemicals such as antifeedants and pheromones.

It would certainly be highly desirable to find potent, broad spectrum antifeedants with systemic activity to protect the new, unsprayed growth. Here we are faced with a bewildering variety of natural compounds of mostly limited potencies and small activity spectra (Schoonhoven 1982). In order to be reliable an antifeedant should prevent starving larvae from eating their hostplants. Thus far few compounds which are easily accessible to synthetic chemistry have met these criteria. Azadirachtin may be very important to local subsistance farming. Yet so far is has not proved to be a particularly attractive lead for the synthetic chemist.

The pheromones, again a group of great structural variety and high specificity, have struggled for two decades now for a niche in the market. Manipulating the behaviour of male adult insects has not proved a reliable means to prevent crop losses which are mostly due to larval feeding.

I believe that pheromones will probably find a small niche in special applications, in population monitoring or as some sort of synergists in a lure and kill approach. But they certainly will not replace insecticides.

Again, I would like to emphasize that in terms of progeny and impact only the second generation of insecticides has been proliferating substantially. This group alone has achieved great commercial success and today these neurotoxic molecules are still the basis for reliable insect control operations.

I conclude that at present we see no realistic alternative to insecticidal potency. We will have to watch out, however, for selectivity and environmental degradability.

In my view the real challenge lies in the fact that, thus far, only few neurochemical targets have been exploited for insecticidal activity. Yet insect neurochemistry provides several, well-described, highly promising biochemical systems such as the postsynaptic proteins triggered by the aminoacids glutamate and GABA, the biogenic amines or the only just emerging class of neuropeptides. Obviously this volume is focused largely on these aspects of insect neurochemistry.

3 The Maturity of the Market

A brief look into either the ecology or the history of agriculture will tell us that man has always been and will always be competing with other herbivores for his crops. We simply have to accept the fact that we always will have to defend our agricultural products against our herbivorous competitors in some way.

It is our business to provide the farmer with means to increase the productivity of his efforts. That is to minimize his inputs in terms of labour and capital and to maximize his rewards in terms of yield and profit. We must also provide international organizations such as WHO with means to control desease vectors, which can be so detrimental to man and his livestock.

Yet the arena for this scenario is not static; we are witnessing a fascinating speed in the evolution of insect populations. The first signs of resistance can sometimes be demonstrated after only a few generations under continuous, strong selection pressure (Sawicki et al. 1984).

Today, there is a large number of resistant insect populations which can no longer be kept in check by the available insecticides (Georghiou and Mellon 1983).

The agricultural markets in the world are a fluctuating and evolving environment. Crop varieties change. Notorious pest problems emerge and disappear. Agricultural policies change. The public opinion about the desirability of our products and consequently the registration requirements have been changing substantially and will continue to do so in the future. The withdrawal of several chlorinated hydrocarbons and the subsequent introduction of the synthetic pyrethroids is a clear illustration of such changes. Another, very important aspect has to be mentioned in this context. The fast-growing human populations in most developing countries require much more food than is currently available to them. We have to help to increase the agricultural productivity in those communities.

4 Conclusion

The challenges in our profession are numerous. It is our main objective to improve the desired activity of our compounds against the target species and at the same time to minimize possible, undesirable side effects. This is an area of continuing, technological advance comparable to other branches of modern industry.

We are producing compounds for an evolving and fluctuating market. Yet the molecules we are synthesizing today will hardly reach the market before 1990. They will be commercial products during the first decades of the next century.

It is my personal conviction that the development of new insecticides for the coming decades is not just a formidable scientific challenge but a real obligation deserving all our enthusiasm and skill.

References

Barbers RH, Pratt JJ (1951) Development of insect resistance to insecticides. II. A critical review of the literature up to 1951. US Dep Agric Agric Res Administ Bur Entomol Plant Quarant Rep E-818

Forgash AJ (1984) History, evolution and consequences of insecticides resistance. Pestic Biochem Physiol 22:178–186

Geissbühler H (1984) Biorational reflections in agricultural chemical research. Chimia 38:307–361

Georghiou GP, Mellon RB (1983) Pesticide resistance in time and space. In: Georghiou GP, Saito T (eds) Pest resistance to pesticides. Plenum Press, New York London, pp 1–46

Hummel HE (1983) Insecticides and their design. J Nematol 15:615–639

Magee PS, Cohn GK, Menn JJ (1984) Pesticide synthesis through rational approaches. Am Chem Soc, Washington D C

Melander AL (1914) Can insects become resistant to sprays? J Econ Entomol 7:167

Sawicki RM, Denholm J, Farnham AW, Devonshire AL (1984) Recent research into resistance to pyrethroid insecticides. Paper Present 17th Int Congr Entomol, Hamburg, Aug 20–26

Schoonhoven LM (1982) Biological aspects of antifeedants. Entomol Exp Appl 31:57–69

The Biochemical Mode of Action of Insecticides [1]

A. C. BAILLIE [2]

1 Introduction

The aim of this paper is to give a brief review of the interactions of insecticides with their target sites. Insecticides interfere with energy production, cuticle deposition, hormone action and the nervous system, and these areas will be discussed in turn.

2 Respiration

Respiration is composed of a number of different steps, many of which are affected by insecticides. Organic foodstuffs contain mixtures of carbohydrates, fats and proteins and these are broken down to sugars, acetate (as acetyl CoA) and aminoacids, respectively. Sugars are then converted by glycolysis to pyruvic acid, an intermediate which is also accessible from fats (via glycerol) and aminoacids. The reactions of glycolysis take place in the soluble phase of the cell and the pyruvate is subsequently transported into the mitochondrion where it is converted by the pyruvate dehydrogenase multi-enzyme complex to acetyl CoA. The acetyl CoA formed is then broken down further in the tricarboxylic acid (TCA) cycle. The pyruvate dehydrogenase complex and the analogous α-oxoglutarate dehydrogenase complex which occurs in the TCA cycle consist of three enzymes and use a number of cofactors. One of these is lipoic acid which is reduced to the corresponding dithiol during the reaction sequence, and it is likely that arsenic-containing compounds produce an arsenic species which inhibits enzyme activity by combining with this dithiol (and probably other cellular dithiols as well) (Webb 1966). It has recently been proposed that the actual toxicant is arsenous acid, $HAs(OH)_2$, and that this is formed by direct reduction of arsenites and by a two-step reduction of arsenates (Knowles and Benson 1983).

An alternative possibility for the arsenate (As^V) ion is that it mimicks a phosphate ion and becomes incorporated into key high-energy containing intermediates which rapidly break down again (Slater 1963). The net result is the futile recycling of intermediates instead of a supply of the required phosphorylated products.

1 This paper was presented at the meeting by J. R. Corbett. It is overlapping in part with a paper given by J. R. Corbett at the joint SCI/RSC International Symposium on Recent Advances in the Chemistry of Insect Control (Cambridge, 1984) which has been published by the Royal Society of Chemistry

2 FBC Limited, Chesterford Park Research Station, Saffron Walden, Essex, CB10 1XL, UK

Approaches to New Leads for Insecticides
Edited by von Keyserlingk et al.

In the course of the oxidative reactions of respiration the cofactor NAD^+ becomes reduced to NADH. In order to continue to fulfill its role, therefore, the cofactor has to be regenerated. Thus, in a series of reactions, NADH is reoxidized by relinquishing its electrons through a chain of carriers to oxygen, which is reduced to water.

The insecticide rotenone whose actions are described in more detail in Fukami's chapter, inhibits this electron-transport process particularly strongly at an early step (Horgan et al. 1968), but it also has effects elsewhere on the carrier chain (Horgan et al. 1968) and on cell division (Marshall and Himes 1978). Cyanide, which is used as a fumigant, also inhibits electron-transport, but at the final step, that catalyzed by cytochrome oxidase (Dixon and Webb 1964). Cyanide probably combines with essential iron and copper atoms in the active site of the enzyme, but the action is not specific since many other enzymes are also sensitive to the cyanide ion (Dixon and Webb 1964).

During the flow of electrons down the respiratory electron-transport chain, energy is trapped by the synthesis of the high-energy intermediate ATP from ADP and inorganic phosphate, in a process called oxidative phosphorylation. The mechanism of the coupling of ATP synthesis to electron transport is a subject of continuing research and debate but most biochemists favour the "chemiosmotic" theory (Mitchell 1961) which states that, during the passage of electrons down the chain of respiratory carriers, protons are transported across the inner mitochondrial membrane from inside to out. The resulting proton electrochemical gradient drives ATP synthesis via an ATPase enzyme which spans the membrane. (The enzyme is called an ATPase because, in the reverse direction, it can hydrolyse ATP.) Insecticides (and acaricides) interfere with oxidative phosphorylation in one of two ways.

Compounds of the first class, including for example DNOC and other dinitrophenol derivatives, uncouple ATP synthesis from electron transport i.e. electron flow occurs, but ATP is not produced. Uncouplers are typically lipid soluble acids and, on the basis of the "chemiosmotic" theory, it is supposed that they act by destroying the proton gradient referred to above (Mitchell and Moyle 1967). Thus, the protonated acid enters the mitochondrion and dissociates. Since the resulting anion can effectively delocalize its charge, it is deemed able to recross the membrane, pick up another proton, carry it in, and so on. One interpretation of this scenario is that the uncoupler needs no specific binding site, but this has been called into question lately and a specific site has been proposed (Hatefi 1980).

The second class of compounds to affect oxidative phosphorylation includes acaricides such as trialkyl tin compounds (Aldridge et al. 1977, Desaiah et al. 1973), and diaryl sulphides and related compounds (Desaiah et al. 1972). It is thought that they inhibit the ATPase directly although the molecular details of how this occurs are not known. The effect is to inhibit both electron transport and ATP synthesis.

3 Cuticle Deposition

Chitin, which is essentially a polymer of *N*-acetylglucosamine, is a structural component of the insect cuticle essential for proper insect development. It has been known for some time that the insect growth regulator diflubenzuron interferes with larval cuticle deposition (Mulder and Gijswijt 1973) and disrupts the moulting process, and a series of studies have led to the conclusion that the compound inhibits some aspect of chitin synthesis (Hajjar and Casida 1979). However, chitin synthase, the enzyme which catalyzes the final step in chitin synthesis and which has been considered to be the site of action of the compound, is not affected in cell-free systems (Cohen and Casida (1982). The primary effect is therefore still in doubt.

One possibility is that DNA synthesis is inhibited in cells which give rise to the adult epidermis (DeLoach et al. 1981), but the effect may also occur in other types of cells (Soltani et al. 1984). A second suggestion (Marks et al. 1982) is that the activity of the benzoylphenylureas may be explained by an inhibition of chymotrypsin-like proteases involved in the proteoloytic activation of a chitin synthase zymogen.

4 Hormone Action

Considerable effort has recently gone into the search for new insecticides interfering with insect hormone action as described in detail in Menn's chapter. Thus, the discovery of the insect juvenile hormones, whose function is to maintain larvae in a juvenile stage at a moult, opened the way for the synthesis and testing of a variety of synthetic analogues. The presence of such compounds in the insect when the natural hormone level is low would be expected to disturb normal morphogenesis to the adult. So far the search has led to the discovery and development of the juvenile hormone analogues methoprene (Henrick et al. 1973, Bowers 1982) and MV-678 (Schwarz et al. 1974). A limitation of this strategy, however, is that juvenile hormone levels are high for most of the insect life cycle so that synthetic analogues will only disrupt development at limited stages in the cycle. Bowers and his colleagues, therefore, began a search for chemicals which might interfere with the synthesis, movement or action of juvenile hormones, a search which led to the discovery of the precocenes, compounds which interfere with hormone biosynthesis (Bowers et al. 1976, Pratt and Bowers 1977). As to how this might occur, it has been suggested that the double-bond in precocenes is epoxidized, possibly by the same enzyme as catalyzes the epoxidation step in juvenile hormone synthesis, and that the epoxide then alkylates critical sites in the cell (Pratt et al. 1980). Menn concentrates in his chapter on various compounds with antijuvenile hormone activity.

5 The Nervous System

The insect nervous system is by far the most important target site for insecticides. The operation of the nervous system (Shankland 1976, Dowson 1977) and the effects of insecticides on it (Corbett et al. 1984) have recently been reviewed and in this section we briefly indicate where, and if possible how, insecticides interfere.

The discussion will be restricted to chemicals in present use and will not attempt to explore the effects of other chemicals known to interact with the various processes.

As far as we know, no insecticide acts specifically by interfering with sensory perception, although pheromones, which do find limited use in insect control, obviously act here.

On the other hand there is a great deal of evidence to suggest that transmission in sensory, and other neurones, is affected by insecticides, particularly the pyrethroids and DDT. Much of the work has been electrophysiological, but biochemical studies are now beginning to appear. Electrophysiological studies allow investigation of the effects of chemicals on the movement of ions into and out of nerve cells in isolated tissue preparations, and many research groups have shown that DDT and pyrethroids interfere with the closing of sodium channels which have been opened in the course of normal nervous activity (Narahashi 1982, Vijverberg et al. 1982b; see in particular the next chapter in this volume by Miller). This means that, at any particular point on the axon membrane, sodium ions are still moving into the neurone when an action potential would normally have passed and the sodium channels are supposed to be closed. Prolonged sodium tail currents can be recorded experimentally which are capable of generating further action potentials. This is thought to explain the repetitive bursts of action potentials observed in treated tissue from many sources. These symptoms are referred to as Type I symptoms.

Sensory neurones appear to be particularly affected (Gammon et al. 1981) and there is a good correlation between the effects of pyrethroids on sensory neurones and the action of the compounds as knockdown agents (Miller and Adams 1977). This is not surprising since the sensory system could be particularly susceptible to contact insecticides of this type. However, there is no good correlation between the ability of pyrethroids to knock insects down and their ability to kill them (Miller and Adams 1977), so it has been difficult to explain the lethal action of pyrethroids solely on the basis of their effects on sensory neurones. The picture became more confused when the recently discovered cyanopyrethroids were examined in electrophysiological experiments, because they do not induce repetitive activity in sensory neurones. Instead they lead to a slow depolarization of the axonal membrane and neuromuscular block (Type II symptoms) (Narahashi 1982). The search for an explanation of these phenomena is now under way.

Miller's group (Salgado et al. 1983a) has proposed that Type II symptoms are a result of the compounds affecting sodium channels at the pre-synaptic terminal of the neuromuscular junction. The channels are still prevented from closing, but the resulting depolarization leads in this case to enhanced transmitter release and eventually to failure of neuromuscular transmission due to depletion of transmitter. These workers also showed (Salgado et al. 1983b) that DDT and Type I pyrethroids (those causing repetitive activity in sensory nerves) do also affect the neuromuscular junction, an effect which had apparently been overlooked given the dramatic nature of the repetitive firing response. They found that structure-activity correlations and the temperature-dependence of the effect (which we will discuss later) were consistent with toxicity being due to the effect on sodium channels at the nerve terminal.

Narahashi and his colleagues have concluded that the Type I/Type II difference is a reflection of the rates at which the various insecticides bind to the sodium channel (Lund and Narahashi 1983). Type I compounds and DDT are considered to modify the channel quickly, so that the negative after-potential builds up quickly and repetitive firing occurs; such compounds only keep the channels open for a short time before diffusing off. Type II compounds on the other hand affect channels more slowly, but the modified channels also return to normal extremely slowly so that, over a period, there is a slow, but gradually increasing depolarization. Under these conditions repetitive activity does not arise and the increasing depolarization leads to a failure of neurotransmission. Van den Bercken's group has reached similar conclusions (Vijverberg et al. 1982 a) and further clarification of this area can be expected.

Since, with some exceptions to be discussed later, there is broad agreement on the conclusion that these compounds bind to the sodium channel a reasonable question to ask is "Why hasn't anybody isolated a population of channels and examined their interaction with the chemicals?" The fact is, however, that such experiments are very difficult to do and so such studies are in their infancy. Nevertheless a start has been made using classical biochemical techniques. Thus, Chang and Plapp (1983) have shown the existence, in housefly brain, of a membrane receptor which specifically binds radiolabelled DDT and cis-permethrin. DDT reduced the extent of binding of, for example, cis-permethrin and vice versa, indicating that both compounds cannot occupy the receptor at the same time.

Soderlund et al. (1983) have found similar stereospecific binding of a pyrethroid to putative receptors from mouse brain membranes. The confirmation and extension of such studies should be of great interest, and could put on a firm basis earlier proposals by, for example, Holan (1969) on the existence of a specific receptor for DDT.

The interaction between these insecticides and a putative receptor bears on an intriguing property of DDT and the pyrethroids, namely that they are more insecticidal at lower temperatures. This property, which is not due to reduced metabolism at low temperatures, has invariably been addressed in any neurophysiological study. In some but not all of these, the repetitive activity induced in sensory neurones is more pronounced at lower temperatures, and the depolarizing effect in the central nervous system and at nerve terminals is also increased at lower temperatures. These effects are usually explained in terms of the receptor-insecticide complex dissociating more slowly at lower temperatures (Holan 1969). If one adds to this the possibility that the sodium channels close more slowly at lower temperatures (Van den Bercken et al. 1979) it can be seen that the overall result would be a particularly prolonged sodium ion influx and thus enhanced neurotoxicity.

Before leaving DDT and the pyrethroids mention should be made of alternative proposals on their mode of action which are most conveniently considered here. Regarding DDT, Matsumura and his colleagues have repeatedly advanced the idea that the compound interferes with the regulation of calcium levels at the surface of the axonal membrane, in such a way as to produce repetitive activity (Ghiasuddin and Matsumura 1979). Although this proposal cannot be ruled out, corroborating evidence is still needed.

Finally Lawrence and Casida (1983) have proposed that Type II pyrethroids in some way antagonize the action of GABA, at the GABA-receptor/ion channel complex. At first sight it seems surprising that the presence of the cyano group could move the binding site from the sodium channel to the GABA receptor and/or its associated chloride channel and further elucidation of this effect promises to be of great interest. In this connection it should be noted that the action of DDT and pyrethroids is not absolutely unique to sodium channels since the compounds increase the electrical activity of neurosecretory neurones in which the inward current is carried largely by calcium ions, rather than sodium ions (Orchard and Osborne 1979). The resulting, inappropriate hormone release would, of course, contribute to the overall poisoning process.

Let us turn now to how insecticides interfere with the events associated with synaptic transmission in the central nervous system. There is good evidence that dieldrin (Shankland 1979) (and probably other cyclodienes and related halogenated compounds) (Bloomquist and Shankland 1983) and γ-HCH (Uchida et al. 1978) cause an excessive release of transmitter from nerve terminals. The cause of this is not clear, but one possibility, applicable to transmitter release generally, arises from studies on heptachlorepoxide, which may be the toxic species in the action of the cyclodiene insecticide, heptachlor. Thus, Yamaguchi et al. (1980) on the basis of results with rat brain tissue, concluded that the chemical led to increased levels of calcium ions in the nerve terminal by interfering either with the calcium pump on the pre-synaptic membrane or calcium sequestration by endoplasmic reticulum. The increased levels of calcium in the nerve terminal would lead to increased transmitter release. More work is required, however, to convincingly correlate insecticidal activity of γ-HCH, and dieldrin and its relatives, with effects on calcium transport systems.

An alternative suggestion, which will be discussed in more detail in the chapters by Miller and by Eldefrawi, is that the cyclodienes interfere with the proper functioning of inhibitory synapses, at which GABA is the neurotransmitter. Such an effect could, of course, account for the hyperexcitation seen in treated insects (Tanaka et al. 1984). Next we will consider what is probably the best characterized site of insecticide action, interference with the removal of acetylcholine from the synapse by the acetylcholinesterase-inhibiting organophosphorus compounds and carbamates. Although organophosphorus compounds, in particular after biological oxidation, are reactive compounds with the capacity to react with a number of enzymes, the prevailing hypothesis is still that acetylcholinesterase is their primary site of action. Although it has recently been reported that some of these compounds can also bind to the acetylcholine receptor with high affinity (Eldefrawi and Eldefrawi 1983) we will focus here on the classical concept of acetylcholinesterase inhibition.

Acetylcholinesterase can be purified readily and the effects of inhibitors examined in detail. The mechanism of action of the enzyme both with acetylcholine and with inhibitors is well understood in molecular detail. The active site contains a serine residue and, in the physiological reaction, this attacks the ester carbonyl group of acetylcholine and becomes acetylated. The acetyl enzyme is then rapidly cleaved in a second step to regenerate the free enzyme. Organophosphorus and carbamate insecticides also react with the serine to give a phosphorylated or carbamylated enzyme, so that serine is effectively masked, the enzyme is inhibited and acetylcholine accummulates in the synapse.

There are two distinct steps in the inhibition reaction (Main 1964), initial reversible formation of an enzyme-inhibitor complex, followed by phosphorylation or carbamylation. The factors governing these have been examined in some detail and it seems clear that the inhibitory potency of carbamates is determined mainly by how well they complex to the enzyme (Kamoshita et al. 1979), whereas that of organophosphorus compounds is usually determined mainly by their phosphorylation rate (O'Brien 1976).

As indicated above, insecticides are also known to act at the acetylcholine receptor. Although the receptor from electric fish has been particularly well studied (Maelicke 1984, Verdenhalven and Hucho, this volume), an increasing amount of information is available from insect tissue (Sattelle 1980). More details on this aspect are described in Breer's chapter. Electrophysiological studies and experiments with cell-free receptor preparations have led to the conclusion that, at low concentrations, nicotine binds to the acetylcholine receptor and induces the typical depolarizing response i.e. it acts as an agonist (Sattelle 1980, Gepner et al. 1978).

Whereas nicotine binds so as to induce depolarization, the main action of the natural product nereistoxin and of the related insecticide cartap, which probably gives rise to nereistoxin in vivo, is that they bind but do not induce depolarization i.e. they are antagonists (Callec et al. 1980, Eldefrawi et al. 1980). Experiments using a number of (non-insect) preparations suggested that these compounds bind at the acetylcholine site rather than elsewhere on the receptor-ion channel complex (Eldefrawi et al. 1980), but it is not yet certain that this is so in insects.

As mentioned earlier in the discussion of the mode of action of the cyclodienes, yet another synaptic site has been suggested as a target for insecticides, the GABA receptor at inhibitory synapses in the peripheral nervous system of insects. Evidence is accumulating that the recently discovered avermectins affect neuromuscular function. In particular, avermectin B1a appears to interfere in some way with the action of GABA and/or its associated chloride channel at inhibitory synapses (Campbell et al. 1983). The compound is thought to bind to a site on the complex which is different from the GABA-binding site, but acts to enhance binding of GABA to its site (Wang and Pong 1982).

Finally we will briefly mention the effects of insecticides on the action of octopamine, the neurophysiological role of which has begun to emerge in recent years (Leake and Walker 1980). The compound appears to act as a transmitter at some synapses, while in other cases it exerts a modulatory effect. There is now good evidence that formamidine insecticides are metabolised to give derivatives which act as agonists at octopamine receptors. For further details see Evans' chapter on biogenic amines.

6 Concluding Remarks

In concluding this introduction to the subject three points are worth emphasising.

1. Most insecticides have been discovered and developed in the absence of detailed knowledge of their molecular mode of action.
2. The exploration of the mode of action of a particular class of compounds may require many years of intensive research and the emerging picture often contains contradictions, because initial in vitro studies frequently demonstrate that the insecticide is capable of interfering with several biochemical systems. Subsequently, different aspects may be emphasised by different researchers. Clarification of the primary site of action is, therefore, very important and work in this area deserves continuing support.
3. The concept of an alternative approach to insecticide discovery, starting with the target process and utilising in vitro systems is attractive. This approach requires extrapolations and predictions from previous findings, but these have a good chance of being reliable in a situation where only a few variables influence the results. Therefore, good in vitro activity against the chosen target can often be obtained.

   However, many factors other than intrinsic activity in vitro have an impact on the development of a compound for use in the field. These include the costs of synthesis, environmental stability, mammalian toxicity, behaviour on and inside treated plants, uptake, metabolism, excretion and tissue penetration. Attempts to embark on the biochemical design of insecticides should be undertaken with an awareness of these. Nevertheless, we are convinced that efforts to design bioactive molecules represent a valuable adjunct to conventional chemical synthesis.

Acknowledgement. The author is grateful to Dr. D. B. Sattelle of the A.F.R.C. Unit of Insect Neurophysiology and Pharmacology, Cambridge, for his helpful comments on this manuscript.

References

Aldridge WN, Street BW, Skilleter DN (1977) Halide-dependent halide-independent effects of triorganotin and triorganolead compounds on mitochondrial functions. Biochem J 168:353–364

Bercken J van den, Kroese ABA, Akkermans LMA (1979) Effects of insecticides on the sensory nervous system. In: Narahashi T (ed) Neurotoxicology of insecticides and pheromones. Plenum Press, New York London, pp 183–210

Bloomquist JR, Shankland DL (1983) The mode of action and neurotoxicity of mirex, chlordecone and four hydrogenated mirex analogs. Pestic Biochem Physiol 19:235–242

Bowers WS (1982) Endocrine strategies for insect control. Entomol Exp Appl 31:3–14

Bowers WS, Ohta T, Cleere JS, Marsella PA (1976) Discovery of insect antijuvenile hormones in plants. Science 193:542–547

Callec JJ, Sattelle DB, Hue B, Pelhate M (1980) Central synaptic actions of pharmacological agents in insects: Oil-gap and mannitol-gap studies. In: Insect neurobiology and pesticide action (Neurotox 79). Proc Symp Soc Chem Indust Symp, pp 93–100

Campbell WC, Fisher MH, Stapley EO, Albers-Schönberg G, Jacob TA (1983) Ivermectin: A potent new antiparasitic agent. Science 221:823–828

Chang CP, Plapp FW Jr (1983) DDT and pyrethroids: Receptor binding and mode of action in the housefly. Pestic Biochem Physiol 20:76–85

Cohen E, Casida JE (1982) Properties and inhibition of insect integumental chitin synthetase. Pestic Biochem Physiol 17:301–306

Corbett JR, Wright K, Baillie AC (1984) The biochemical mode of action of pesticides. Academic Press, London New York

Desaiah D, Cutkomp LK, Koch RB, Yap HH (1972) Tetradifon (Tedion®): A specific inhibitor of Mg^{2+} dependent mitochondrial adenosine triphosphatase activity. Life Sci 11:389–395

Desaiah D, Cutkomp LK, Koch RB (1973) Inhibition of spider mite ATPases by plictran and three organochlorine acaricides. Life Sci 13:1693–1703

Dixon M, Webb EC (1964) Enzyme inhibitors. In: Enzymes. Longman, London, pp 315–358

Dowson RJ (1977) An introduction to the principles of neurophysiology. Pestic Sci 8:651–660

Duran A, Cabib E (1978) Solubilisation and partial purification of yeast chitin synthetase. Confirmation of the zymogenic nature of the enzyme. J Biol Chem 253:4419–4423

Eldefrawi AT, Bakry NM, Eldefrawi ME, Tsai MC, Albuquerque EX (1980) Nereistoxin interaction with the acetylcholine receptor ionic channel complex. Mol Pharmacol 17:172–179

Eldefrawi ME, Eldefrawi AT (1983) Neurotransmitter receptors as targets for pesticides. J Environ Sci Health B 18(1):65–88

Gammon DW, Brown MA, Casida JE (1981) Two classes of pyrethroid action in the cockroach. Pestic Biochem Physiol 15:181–191

Gepner JI, Hall LM, Sattelle DB (1978) Insect acetylcholine receptors as a site of insecticide action. Nature (London) 276:188–190

Ghiasuddin SM, Matsumura F (1979) DDT inhibition of Ca^{++}-ATPase of the peripheral nerves of the American lobster. Pestic Biochem Physiol 10:151–161

Hajjar NP, Casida JE (1979) Structure activity relationships of benzoylphenyl ureas as toxicants and chitin synthesis inhibitors in *Oncopeltus fasciatus*. Pestic Biochem Physiol 11:33–45

Hatefi (1980) Discovery and photoaffinity labelling of the mitochondrial uncoupler-binding site. Ann NY Acad Sci 346:434–443

Henrick CA, Staal GB, Siddall JB (1973) Alkyl 3,7,11-trimethyl-2,4-dodecadienoates, a new class of potent insect growth regulators with juvenile hormone activity. J Agric Food Chem 21:354–359

Holan G (1969) New Halocyclopropane insecticides and the mode of action of DDT. Nature (London) 221:1025–1029

Horgan DJ, Ohno H, Singer TP, Casida JE (1968) Studies on the respiratory chain-linked reduced nicotinamide adenine dinucleotide dehydrogenase. J Biol Chem 243:5967–5976

Kamoshita K, Ohno I, Fujita T, Nishioka T, Nakajima M (1979) Quantitative structure-activity relationships of phenyl N-methyl-carbamates against house fly and its acetylcholinesterase. Pestic Biochem Physiol 11:83–103

Knowles FC, Benson AA (1983) The biochemistry of arsenic. Trends Biochem Sci 8:178–180

Lawrence LJ, Casida JE (1983) Stereospecific action of pyrethroid insecticides on the γ-aminobutyric acid receptor-ionophore complex. Science 221:1399–1401

Loach JR De, Meola SM, Mayer RT, Thompson JM (1981) Inhibition of DNA synthesis by diflubenzuron in pupae of the stable fly *Stomoxys calcitrans (L)*. Pestic Biochem Physiol 15:172–180

Leake LD, Walker RJ (1980) Invertebrate neuropharmacology. Blackie and Son, Glasgow

Lund AE, Narahashi T (1983) Kinetics of sodium channel modification as the basis for the variation in the nerve membrane effects of pyrethroids and DDT analogs. Pestic Biochem Physiol 20:203–216

Maelicke A (1984) Biochemische Aspekte der cholinergen Reizung. Angew Chem 96:193–219

Main AR (1964) Affinity and phosphorylation constants for the inhibition of esterases by organophosphates. Science 144:992–993

Marks EP, Leighton T, Leighton F (1982) Modes of action of Chitin synthesis inhibitors. In: Coats JR (ed) Insecticide mode of action. Academic Press, London New York, pp 281–313

Marshall LE, Himes RH (1978) Rotenone inhibition of tubulin self-assembly. Biochim Biophys Acta 543:590–594

Miller TA, Adams ME (1977) Central versus peripheral action of pyrethroids on the housefly nervous system. In: Elliott M (ed) Synthetic pyrethroids. Am Chem Soc, Washington DC, pp 98–115

Mitchell P (1961) Coupling of phosphorylation to electron and hydrogen transfer by a chemi-osmotic type of mechanism. Nature 191:144–148

Mitchell P, Moyle J (1967) Acid-base titration across the membrane system of rat liver mitochondria, catalysis by uncouplers. Biochem J 104:588–600
Mulder R, Gijswijt MJ (1973) The laboratory evaluation of two promising new insecticides which interfere with cuticle deposition. Pestic Sci 4:737–745
Narahashi T (1982) Insecticides as probes for the study of ionic channels in nerve membranes. In: Neuropharmacology of insects. Ciba Found Symp 88. Pitman, London, pp 291–306
O'Brien RD (1976) Acetylcholinesterase and its inhibition. In: Wilkinson (ed) Insecticide biochemistry and physiology. Plenum Press, New York London, pp 271–296
Orchard I, Osborne MP (1979) The action of insecticides on neurosecretory neurons in the stick insect *Carausius morosus*. Pestic Biochem Physiol 10:197–202
Pratt GE, Bowers WS (1977) Precocene II inhibits juvenile hormone biosynthesis by cockroach corpora allata in vitro. Nature (London) 265:548–550
Pratt GE, Jennings RC, Hamnett AF, Brooks GT (1980) Lethal metabolism of precocene-I to a reactive epoxide by locust corpora allata. Nature (London) 284:320–323
Salgado VL, Irving SN, Miller TA (1983 a) Depolarisation of motor nerve terminals by pyrethroids in susceptible and kdr-resistant house flies. Pestic Biochem Physiol 20:100–114
Salgado VL, Irving SN, Miller TA (1983 b) The importance of nerve terminal depolarisation in pyrethroid poisoning of insects. Pestic Biochem Physiol 20:169–182
Sattelle DB (1980) Acetylcholine receptors of insects. In: Berridge MJ, Treherne JE, Wigglesworth VB (eds) Advances in insect physiology, vol XV. Academic Press, London New York, pp 215–315
Schwarz M, Miller RW, Wright JE, Chamberlain WF, Hopkins DE (1974) Compounds related to juvenile hormone. Exceptional activity of arylterpenoid compounds in four species of flies. J Econ Entomol 67:598–601
Shankland DL (1976) The nervous system: Comparative physiology and pharmacology. In: Wilkinson CF (ed) Insecticide biochemistry and physiology. Plenum Press, New York, pp 265–270
Shankland DL (1979) Action of dieldrin and related compounds on synaptic transmission. In: Narahashi T (ed) Neurotoxicology of insecticides and pheromones. Plenum Press, New York London, pp 139–153
Slater EC (1963) Uncouplers and inhibitors of oxidative phosphorylation. In: Hochster RM, Quastel JH (eds) Metabolic inhibitors, vol II. Academic Press, London New York, pp 503–516
Soderlund DM, Ghiasuddin SM, Helmuth DW (1983) Receptor-like stereospecific binding of a pyrethroid insecticide to mouse brain membranes. Life Sci 33:261–267
Soltani N, Besson MT, Delachambre J (1984) Effects of diflubenzuron on the pupal-adult development of *Tenebrio molitor* L. (*Coleoptera, Tenebrionidae*): growth and development, cuticle secretion, epidermal cell density, and DNA synthesis. Pestic Biochem Physiol 21:256–264
Tanaka K, Scott JG, Matsumura F (1984) Picrotoxinin receptor in the central nervous system of the American cockroach: its role in the action of Cyclodiene-type insecticides. Pestic Biochem Physiol 22:117–127
Uchida M, Fujita T, Kurihara N, Nakajima M (1978) Toxicities of γ-BHC and related compounds. In: Shankland DL, Hollingworth RM, Smyth T Jr (eds) Pesticide and venom neurotoxicity. Plenum Press, New York London, pp 133–151
Vijverberg HPM, Ruigt GSF, Bercken J van den (1982 a) Structure-related effects of pyrethroid insecticides on the lateral-line sense organ and on peripheral nerves of the clawed frog.l *Xenopus laevis*. Pestic Biochem Physiol 18:315–324
Vijverberg HPM, Zalm JM van der, Bercken J van den (1982 b) Similar mode of action of pyrethroids and DDT on sodium channel gating in myelinated nerves. Nature (London) 295:601–603
Wang CC, Pong SS (1982) Actions of avermectin B1a on GABA nerves. Membranes and genetic disease. Alan R Liss Inc, New York, pp 373–395
Webb JL (1966) Enzyme and metabolic inhibitors, vol III, chap 6. Arsenicals. Academic Press, London New York, pp 595–793
Yamaguchi I, Matsumura F, Kadous AA (1980) Heptachlor epoxide: Effects on calcium mediated transmitter release from brain synaptosomes in rat. Biochem Pharmacol 29:1815–1823

Can Mode of Action Studies Guide Chemical Synthesis?

T. A. MILLER [1]

1 Introduction

Chemical industry has been exploring alternative means of finding new insecticides to supplement traditional screening of chemicals. The need to reexamine the means of finding active compounds was brought about partly by a decreased success rate coming from straightforward bioassay screens (Menn 1983) and partly by the realization that much activity could well have been missed in the past. This is particularly true of behavior-modifying chemicals when screens are designed only to detect toxicity.

Over the past 5 years, major agrochemical firms concerned with insecticides have been hiring young scientists with some training in insect neurobiology or neurotoxicology. I am sure this recent interest in acquiring some expertise in the neurobiology area is directly related to the desire to broaden the methods of searching for chemicals for use in controlling insects. With this in mind, we turn to the title of this chapter and ponder what the role of the neurobiologist might be in agrochemical industry.

To examine the role of mode of action studies in chemical industry, one should keep in mind that besides random screening, two other methods have had good success in finding new insecticides: one is examining natural products for activity and another studying structural analogs of known insecticides. These methods have lead to pyrethroids (from pyrethrins), carbamates (from physostigmine), cartap (from nereistoxin), and arguably avermectin to name a few.

We might also keep in mind that in a practical sense, certain studies cannot or should not be attempted in the context of chemical industry. Instead, they would be conducted in universities or institutes free from marketing pressures. Such an environment allowed Dr. Elliott to discover first bioresmethrin, then deltamethrin, and Dr. B. Brown (with A. Starratt) to isolate and identify the neuropeptide proctolin, both of which took several years; both efforts were protected from termination by men of extraordinary courage and vision, Dr. C. Potter and Dr. E. Spencer, respectively.

Neurobiologists can do several things in chemical industry. They first and foremost contribute knowledge, interpret new research findings from academia and provide an invaluable insight that complements the strictly chemical outlook. One of the most valuable services a neurobiologist can perform for chemical industry is to determine more precisely where a compound is acting (on the nervous

1 Division of Toxicology and Physiology, Department of Entomology, University of California, Riverside, CA 92521, USA

Approaches to New Leads for Insecticides
Edited by von Keyserlingk et al.

system or elsewhere) and what a compound is doing. In short, the neurobiologist can bring a greater focus to the bioassay screens, make the measurement of activity more quantitative and qualitative.

Studies on the mode of action of insecticides can provide new insights into how a chemical is interacting with the integument, how a compound is acting on the target site, if there is lethal synthesis at work, and how much metabolism is affecting a compound, for example. Mode of action studies are most satisfying when one can compare closely related structures at a target site such as the now classic studies on cholinesterase inhibition.

It is a painful truth that despite years of research, studies of the insect nervous system one or more days after poisoning are still rare, and most neurophysiological studies are done by observing the first events in poisoning. This can be misleading. A disruption in water balance has been suggested as one fatal lesion in insects due to insecticide poisoning. More recently, it has become popular to attribute the fatal event even more broadly to neurohormonal dysfunction. Despite some very welcome research in this direction, a description of the fatal events is still missing; possibly one such event would not apply to all insects.

2 Symptoms of Poisoning

We could all use a closer look at the poisoning process itself. Too often chemicals are handed to bioassay personnel and chemists receive back pluses and minuses. An exclusive dependence on this procedure misses a great deal, perhaps even vital information. From the physiological standpoint, a motionless insect could mean anything from muscle blockage up to and including complete central nervous paralysis. Neuromuscular effects, however, represent the starting point in examining the mode of action of insecticides.

All symptoms, whether lack of motion or hyperactivity, are mediated by the skeletal muscles. The logical first procedure is to determine, in order, whether compounds are acting on the muscle, neuromuscular junction, motor axon or elsewhere to produce symptoms.

The housefly shown in Fig. 1 was poisoned with 39 ng of lindane (γ-hexachlorocyclohexane) 24 h before the photograph was taken (Collins et al. 1979). This housefly, however, is perfectly alive. It had its wings pointed in a peculiar direction, it could not fly and it had one leg up. This sort of thing is very common in many kinds of insecticide poisoning. The process of trying to find out what was wrong with this fly, or to gain some knowledge about what the final consequences of poisoning are, is a subject that rightly belongs in the academic sphere. I cannot imagine a person in chemical industry being allowed to figure out why the fly's front leg is held up as a permanent lesion.

Whether symptoms occur after seconds or days, the most logical approach to investigate their cause is to find out if the compounds are doing something to the muscle. That is the simplest thing to do. The next simplest thing to do is to see if there is any effect on the neuromuscular junction. Then the next simplest thing to do in turn is to see if there is an action on the motor axon. If the motor system is functioning normally in early and late stages of poisoning, then one logically examines the central nervous system.

Fig. 1. Adult female housefly treated 24 h previously with an LD_{50} (39 ng) of lindane (γ-hexachlorocyclohexane). Note the wing held down and the left front leg held up. The female was capable of feeding and cleaning, but incapable of flight. (From Collins et al. 1979)

If you are lucky, you will find a compound that is active on the peripheral part of the nervous system, because it makes things much simpler. If you have a compound that is doing something to the central nervous system, it is much more complicated to analyze. Keeping these principles in mind, I would like to describe work that has been done at Riverside for the past 10 years, studying the mode of action of pyrethroid insecticides.

3 Mode of Action of Pyrethroids

We have developed several preparations over the years, some of them are shown here (Fig. 2), to find out exactly what pyrethroids are doing to the insect. When we started, the prevailing theory was that pyrethroids were active on sodium channels in some way and the prevailing explanation of the days was that negative after potential lead to repetitive discharges and something called burst discharge.

The process of looking at this, as I said before, started with the simplest case looking at motor activity in a whole fly. The motor activity can be measured by inserting wire electrodes through the cuticle into muscle groups to record, for example, flight muscle potentials. This procedure was adopted from Peter Usherwood's laboratory going back to the time I spent as his postdoctoral assistant in 1968. This is one of the many things I learned there and was taken from a study that M. Burns was doing. We adapted this sort of technique to look at houseflies and found that it gave us a tremendous insight into the action of compounds.

Figure 3 is a recording that looks like wall paper it is so regular. These are simply muscle potentials recorded from the dorsal longitudinal flight muscles of houseflies. Each letter and each number is a different muscle and the important thing about this trace is that it is simple and extremely regular. The regularity of the pattern is determined by connections of the flight motor neurons and the interneurons in the central nervous system. The potential itself is a muscle potential signalling the arrival of a single nerve impulse at a singly innervated muscle. So here you have measurements of a central pattern generator that determines the

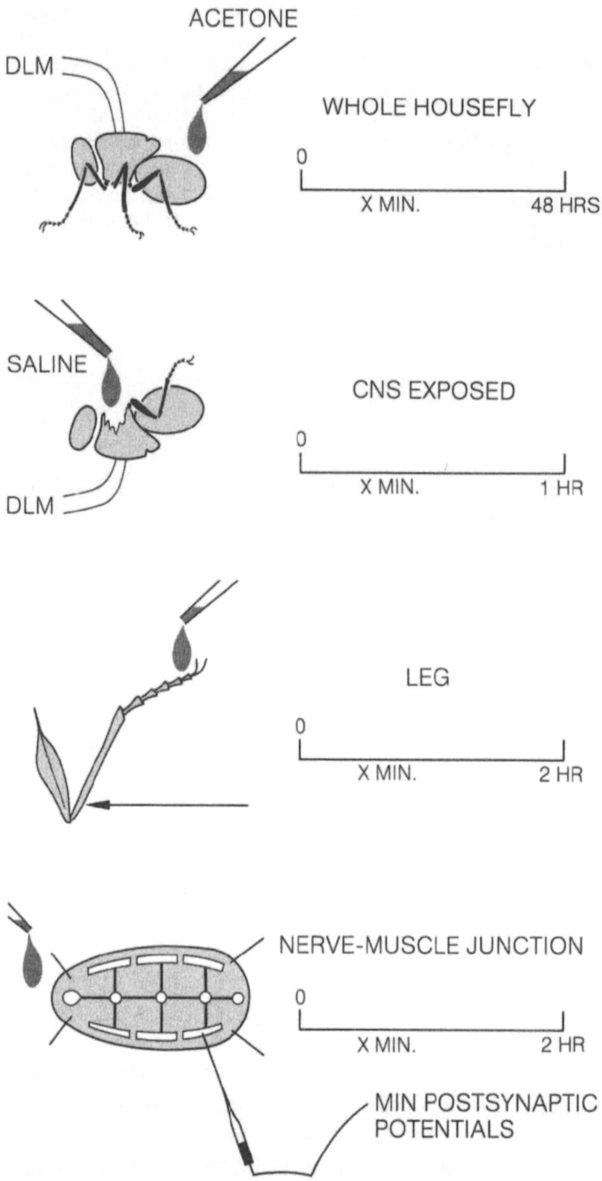

Fig. 2. Methods of analyzing the actions of pyrethroids in houseflies. From *top* to *bottom* they are: (1) whole animal recording of electromyograms, in this case, the dorsolongitudinal flight muscles (*DLM*) and allows the fly to be mounted and examined for several days. (2) Dissecting the inverted adult housefly allows access directly to and application of chemicals on the composite thoracic ganglion by removing the first two pairs of legs. In this case, *DLM*-motor output is monitored as above. (3) Sharpened pins inserted into the leg of the housefly enables sensory nerve impulses to be recorded while chemicals are applied to the tarsi. (4) The larval nerve-muscle preparation allows sophisticated measurements of the actions of chemicals and drugs in the nerve-muscle system. This may also be done with adult muscle preparations, although larval insects are easier to work with. Preparations 2 and 4 require saline, by no means an established procedure for all insects

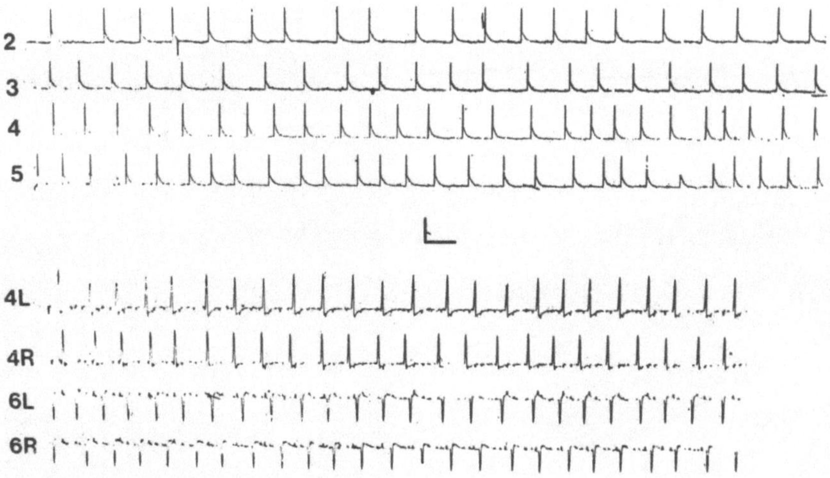

Fig. 3. Recordings from six of the ten motor units in the dorsolongitudinal flight muscle array. The numbers *2, 3, 4* and *5 above* all refer to the right muscles. 4 R is *below* the same muscle as 4 *above* and 6 R *below* is innervated by the same axon that innervates 5 *above*. Muscle potentials are recorded spontaneously in a quiescent adult. The housefly flight motor periodically cycles on producing the pattern shown when the adult housefly is at rest. Since these are fibrillar muscles and are stretch-activated, no contraction occurs and, therefore, no work is done unless and until the wings are unfolded, the pleurosternal muscle contracts, and the thoracic exoskeleton stretches. The antagonistic fibrillar muscle arrays alternately. Calibration 100 ms. (From Adams and Miller 1980)

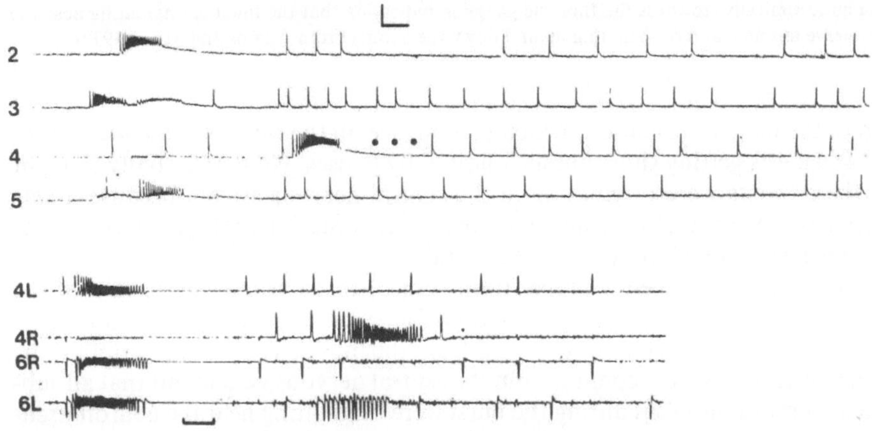

Fig. 4. The houseflies recorded in Fig. 3 were treated with 0.2 μg bioneopynamin less than 10 min before this recording was made. Note the appearance of short bursts of potentials occurring in an otherwise normal recording. Calibration 100 ms. (From Adams and Miller 1980)

order in which the nerve impulses leave the thoracic ganglion to propagate to the peripheral muscles. This preparation, the flight motor recording, was used to great advantage by a student of mine, M. Adams, who worked on this in the late 1970's (Adams 1978). He was looking at, in this case, bioresmethrin poisoning. He found that the beautiful pattern was broken up in the following manner. In-

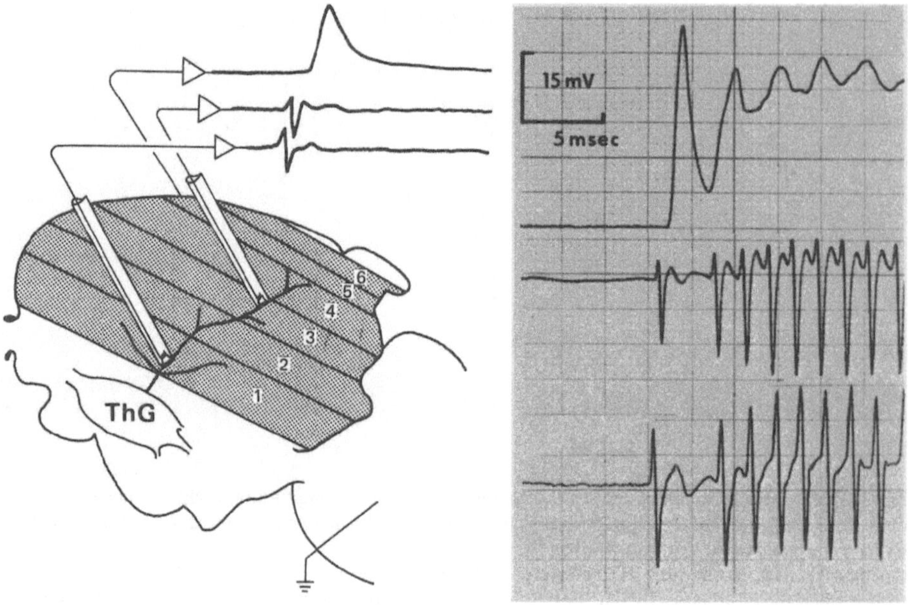

Fig. 5. The poisoning symptoms shown in Fig. 4 were examined using the half-thorax preparation shown. One electrode near the thoracic ganglion (*ThG*) records a nerve impulse propagating orthodromically (towards muscles *5* and *6*) arriving at the distal electrode second and at the muscle third (as indicated on the drawing). All subsequent nerve impulses in the burst shown in the *right* are propagating antidromically (towards the thoracic ganglion indicating that the burst is originating near the motor nerve terminal and propagating up and down the axon. (From Adams and Miller 1979)

stead of the nice coordination, which you can see on the right of these four traces (Fig. 4), he was getting short bursts followed by pauses. Now we initially thought that these bursts were originating in the central nervous system because it was suggested by a group at Rothamsted Experimental Station that all pyrethroid actions were central and none were peripheral.

Dr. Adams developed a preparation (Fig. 5) to determine the origin of the bursts. He recorded from two places along the axon connecting the central nervous system with one flight muscle and from the muscle itself. He found that the first nerve impulses were coming from the central nervous system but that all subsequent nervous impulses during the burst were originating near the neuromuscular junction and propagating both toward the muscle and back toward the thoracic ganglion. Therefore, the poisoning was not actually occurring in the central nervous system; it was occurring in the periphery and bombarding the central nervous system with bursts of nerve impulses propagated antidromically (backwards). Dr. Adams called this peculiar antidromic bursting "backfiring" and was able to call attention for the first time to the peripheral neuromuscular junction as a fairly sensitive site of action.

Dr. Adams did another very interesting thing. This peripheral neuromuscular system in the housefly is very suitable for doing a study, I call the around-the-corner preparation, which was adapted from B. Mulloney's work on *Calliphora*

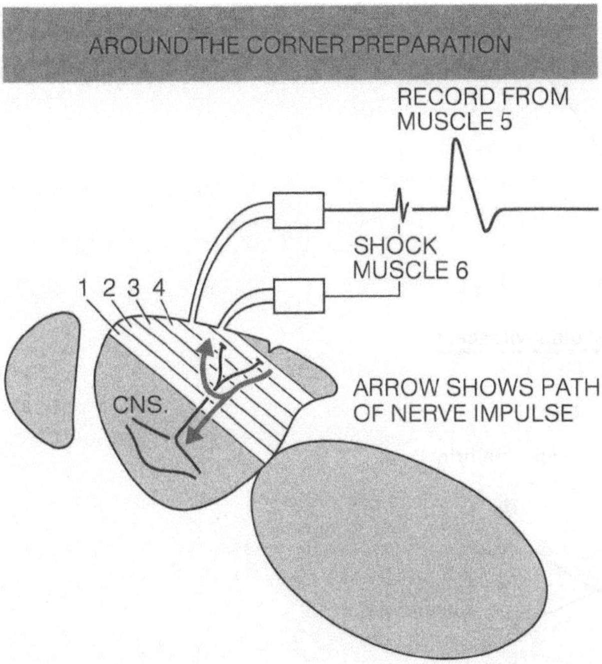

Fig. 6. A drawing showing placement of stimulating electrodes in the sixth (most dorsal) dorsal longitudinal muscle DLM *6* and recording electrodes in the fifth DLM muscle on the same side of the thorax. Stimulation of DLM *6* sends a nerve impulse propagating "around the corner" of the common innervating motor axon to DLM *5* and also antidromically toward the thoracic ganglion where the activity and firing pattern of that one motor neuron is altered, but none of the other motor units are affected. The approach was adapted from a study by Mulloney (1969) on *Calliphora* adults. The adult is shown dissected for clarity, but the preparation was intact adults only

vicina (Mulloney 1969). The dorsal longitudinal flight muscles are singly innervated; however, the top two members of the muscle array, (numbers 5 and 6) happen to be innervated by branches of the same axon, so the 5th axon innervates both the 6th and the 5th muscle. Therefore, if this top muscle is shocked (number 6), a nerve impulse propagates back to the junction and "around the corner" to the 5th muscle plus antidromically back down the trunk of the 5th axon to the CNS (Fig. 6). So in a whole housefly with no dissection at all and no worries about saline, you have a very simple preparation to follow the activity and the function of a given neuromuscular junction. You know which one it is. You know where it is. There is no confusion about what you are recording.

Some pyrethroids, in particular permethrin, have a very high negative temperature coefficient of toxicity. That means that, at room temperature or below, permethrin is fairly toxic but as you raise the temperature permethrin becomes dramatically much less toxic. This is in comparison to fenvalerate and cypermethrin whose characteristics are quite different.

The negative temperature coefficient for permethrin shown in Fig. 7 for *Heliothis virescens* was documented by T. Sparks and his colleagues for the to-

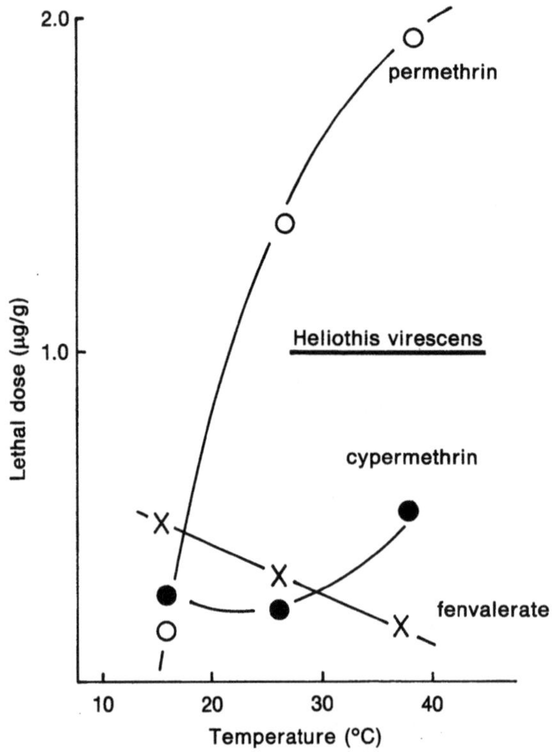

Fig. 7. Graph of the LD_{50} toxicity of three pyrethroids at various temperatures on tobacco budworm larvae, *Heliothis virescens*. Note that above room temperature *permethrin* becomes considerably less toxic and below room temperature slightly more toxic than *fenvalerate*, whose toxicity does not vary as much over 30 °C. Simple substitute of a cyano group alpha to the ester alcohol of *permethrin*, completely changes the temperature vs. toxicity behavior (*cypermethrin*). (From data Sparks et al. 1982)

bacco budworm, but the same sort of response occurs on houseflies. The nice thing about working with pyrethroids and temperature change is that a poisoned insect below room temperature can be converted to an unpoisoned insect by raising the temperature. These conditions are completely reversible within seconds, giving the researcher a tremendous tool.

Dr. Adams altered the temperature during the around-the-corner preparation in *trans*-barthrin-treated adult houseflies (Fig. 8). He found that burst discharges were recorded at room temperature.

If you have never done any study at another temperature, you would think that the burst discharge was characteristic of *trans*-barthrin poisoning, and if the temperature were raised to a region where *trans*-barthrin is less toxic, the burst discharge is lost as expected (Fig. 8). However, when Dr. Adams lowered the temperature below 20 °C, the same thing happened; the burst discharge reverted to single evoked responses (Fig. 8). This surprised us because some of our colleagues had been using burst discharge for the bioassay of pyrethroids in structure-activity studies. Thus, we concluded that burst discharge was an artifact and not critical to poisoning, and brought our attention back to the motor nerve terminal.

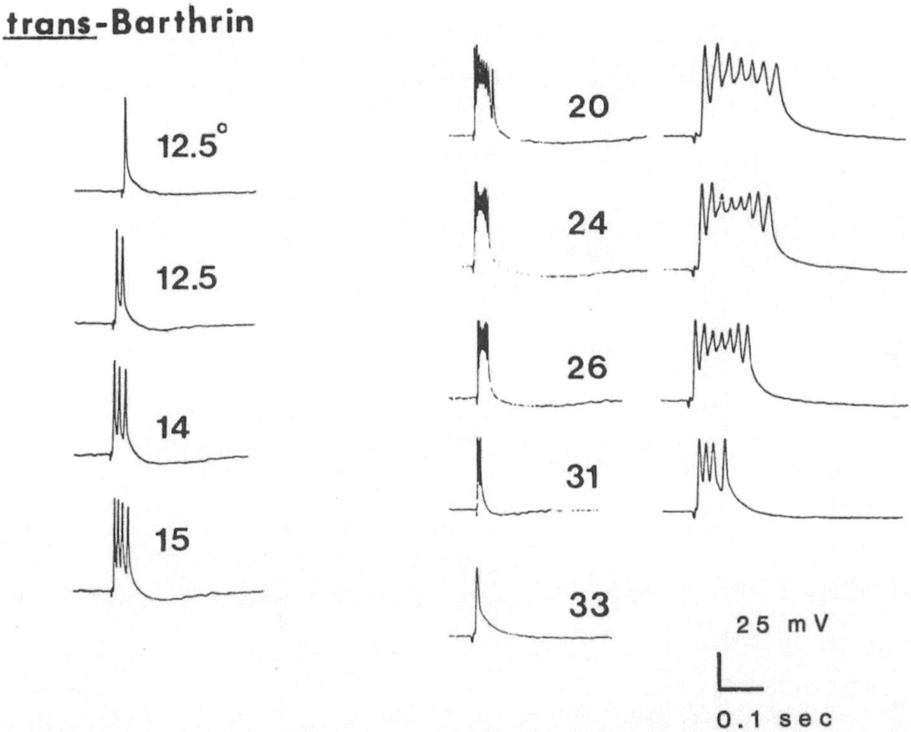

Fig. 8. Muscle potentials recorded from DLM 5 were evoked by stimulation of DLM 6 in the set up shown in Fig. 6. Temperatures were altered by directing hot or cool air at the adult. A thermocouple inserted into the thorax recording the temperatures. The intact adult was topically treated with 1 μg of *trans*-barthrin (about the LD_{50}). Vertical calibration: 25 mV horizontal calibration is 100 ms and 20 ms for the four expanded traces in the *right*. Note the reversion of recorded bursts to single potentials near 12 °C, supposedly the most toxic temperature

The next approach focusing on the neuromuscular junction in studying the mode of action of pyrethroid insecticides was pioneered by S. Irving also in the late 1970's. Dr. Irving came to Riverside as a postdoctoral assistant from the University of Birmingham, and brought with him an interest in the nerve-muscle physiology of dipteran larvae. This preparation revealed itself to be a neurophysiologist's dream. The central nervous system in the cyclorrhaphan diptera is congregated in the front end of the animal including the composite head, thoracic, and abdominal ganglia, and the intersegmental muscles are all innervated by separate nerve bundles that run the length of the animal.

The great advantage of this preparation is that the muscles are single, large flat cells and the neuromuscular junctions are accessible on the surface. This is very unusual for insect muscles, most of which are a series of muscle fibers instead of a single cell layer like this.

I will not describe the elegant physiology that Dr. Irving did while he was at Riverside dealing with neuromuscular transmission and dual innervation (Irving and Miller 1980). I will simply point out that one can record passively from these

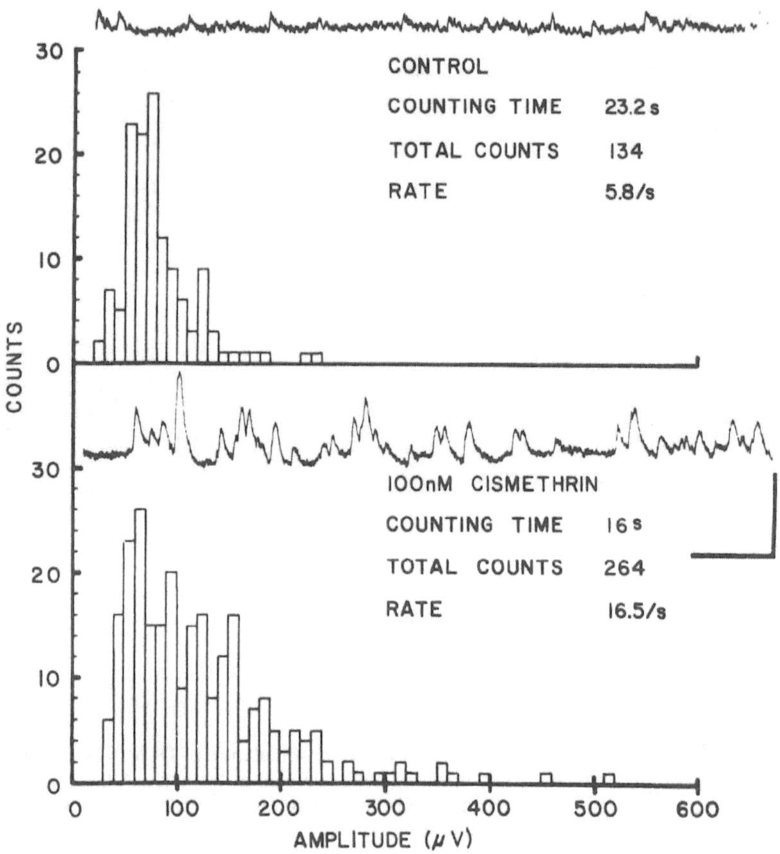

Fig. 9. Amplitude distributions of miniature postsynaptic potentials (mEPSPs) recorded from third instar larvae before (*top*) and after (*bottom*) treatment with 100 nM *cismethrin. Insets* show actual records. Calibration: 0.4 mV, 200 ms. Note the very obvious difference in appearance of recorded mEPSPs. (From Salgado et al. 1983)

flat muscles with a single microelectrode and pick up the spontaneous release of neurotransmitter from the nerve terminal in the form of so-called miniature postsynaptic potentials (mEPSPs) (Fig. 9). To reiterate, these potentials are not evoked. They are recorded at rest, and they provide some idea of the condition of the motor nerve terminal.

Under certain conditions, in particular pyrethroid poisoning, the frequency of the mEPSPs increases greatly and then gradually declines (Fig. 10). This process must have been missed by a large number of previous workers studying the mode of action of insecticides, particularly DDT and pyrethroids. The only reason I can propose why the increase in mEPSP activity was not noticed before is because this process sometimes takes a very long time to develop and, relatively speaking, in the first 30 min of poisoning while the frequency of mEPSPs is increasing, the neuromuscular junction still functions normally even in the face of a great deal of burst discharge activity. When stimulated, a normal post-synaptic potential is

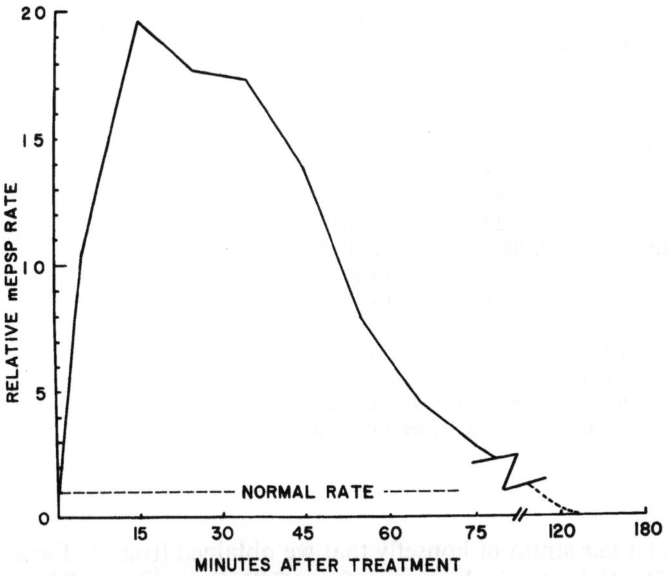

Fig. 10. Plot of relative mEPSP rate recorded from *Musca domestica* larvae at various times after perfusion with 50 nM deltamethrin. (From Miller et al. 1983)

produced and, for all intents and purposes, the muscle and neuromuscular junction are functioning perfectly well during early poisoning. An obvious loss in function comes much later.

We also found out that when the miniature postsynaptic potential activity decreased gradually towards zero later in poisoning, neuromuscular transmission was blocked. This condition, called early block, was followed by a late block at which time the neuromuscular junction was blocked irreversibly. The late block produced a flaccid paralysis of the skeletal muscles of poisoned insects that resembled the middle stage of poisoning (Salgado et al. 1983).

Pyrethroid poisoning, when examined by something as simple as miniature potentials, gave us a very powerful piece of information. It said that if you reached the late block condition, the neuromuscular transmission was irreversibly damaged. This suggested a terminal condition of poisoning more directly related to toxicity in the sense of 24-h mortality than any other. So here was a nerve recording sequence winding up with an irreversible block in a vital function, neuromuscular transmission, and these simple preparations, miniature potentials, muscle recording and so on, lead to a quantitative bioassay (the miniature assay) for pyrethroid insecticides. The bioassay was perfected and developed by S. Irving. Later V. Salgado, doing his Ph. D. at Riverside in the early 1980's, measured the lowest concentration effective in increasing mEPSP rate within 30 min of perfusion.

The first thing we wanted to do with the miniature assay was to check to see whether the miniature assay could detect the presence of the *kdr* gene in pyrethroid-resistant insects. The famous *kdr*-resistance in insects appears to render the nervous system less sensitive to all pyrethroids and DDT. We had three strains

Table 1. Comparison of toxicity and nerve terminal activity in three insect species with a site intensitivity factor. (Taken from Miller et al. 1983)

Insect	Compound	LC_{50} or LD_{50}	EC (M)
Musca S	*cis*-decamethrin	0.6	1×10^{-9}
Musca R	*cis*-decamethrin	8.2	1×10^{-7}
Culex S	*trans*-permethrin	2.1	5×10^{-10}
Culex R	*trans*-permethrin	8,700	1×10^{-6}
Anopheles S	*cis*-permethrin	3.5	5×10^{-11}
Anopheles R	*cis*-permethrin	82.0	1×10^{-9}

Musca domestica NAIDM (S) or 538ge (R); *Culex quinquefasciatus*; *Anopheles stephensi*. All compounds (IR) and (R,S)-α-cyano. EC = lowest concentration effective in increasing mEPSP rate within 1 h. Toxicity in ng insect^{-1} or ppm. S and R, susceptible and resistant, respectively

that were available: (1) a *kdr* strain of housefly that we obtained from A. Farnham at Rothamsted, (2) a *Culex* strain that we got from G. P. Georghiou at Riverside and (3) *Anopheles stephensi,* that he had as well. In fact at this time, G. P. Georghiou was in the process of trying to prove what the resistance factors were in his mosquito strains. He had shown that a high level of resistance to pyrethroids in *Culex quinquefasciatus* was not due to metabolism or penetration factors because it could not be synergized (Omer et al. 1980).

When Dr. Irving used his miniature assay, he found nearly a log 2 molar difference in the standard response in the miniature assay for the *kdr* strain with only a tenfold difference in the toxicity compared to the susceptible strain (Table 1). So the results from the miniature assay showed a 10^2 difference for *Musca*, almost 10^4 for *Culex* and roughly 10^2 for *Anopheles* and were correlated with toxicity differences, and yes indeed, this indicated quite a difference in the site of action, or site insensitivity. So these assays gave us a fairly decent measurement of the intrinsic activity of pyrethroids besides a sophisticated measure of the phenotype of the gene responsible for *kdr*-resistance.

Next I thought the miniature assay technique could be used in an applied way. We tried to see how much *kdr*-resistance was present in several of California's pest insects, but we ran into one problem. The miniature assay takes an hour or two, so one can do no more than a few a day and that is impractical when looking for a resistance factor in the gene pool that is present at the rate of 1 or 2%. It would be better to analyze hundreds of animals a day, so for the past 5 years, we have been trying to come up with faster ways of measuring for the presence of the *kdr*-factor.

One of the first articles referring to knockdown was that by Busvine (1951), in which he described the response of three strains of housefly to DDT. The females of Busvine's susceptible "Rome" strain were 93% knocked down 40 min after treatment (Table 2). In one resistant "Italian" strain (with *kdr*-resistance) 11% were knocked down, and in the other resistant "Sardinian" strain without *kdr*-resistance 80% were knocked down 40 min after treatment. Since both resis-

Table 2. Knockdown of houseflies from three strains after exposure to DDT. (From Busvine 1951)

Percent Knockdown	Susceptible Rome		Resistant Italian		Resistant Sardinian	
After 20 min	86	35	7	0	69	27
After 40 min	100	93	24	11	93	80
% Mortality at 24 h	100	100	19	8	30	11

tant strains wound up with roughly the same LD_{50} (8 and 11%) compared to the susceptible (100% mortality), this meant that the strain with the *kdr* factor could be separated by looking at the symptoms during the first 20–40 min of poisoning. The housefly strain lacking the *kdr* factor behaved like poisoning in the susceptible strain.

This very simple observation led us to try to set up a simple assay of *kdr* in houseflies. We put a mixed population of 40 susceptible and 40 *kdr* adult females in a tower immediately after they were sprayed as a group with a pyrethroid. After 20 min, the 38% of the initial 80 remaining in the tower were all *kdr*. The difference between the susceptible and the *kdr* strain shown is only about tenfold in terms of toxicity. So this is a fairly quick way to separate out *kdr* and susceptible houseflies. Some *kdr* flies come down with the susceptibles, but 100% separation of the *kdr* was possible because of the mixed group in the collection bottle, only the *kdr*'s were active.

This nice assay does not work with adult moths, so it is a bit specialized to something that is flying and fairly active. We needed something a little bit simpler and something that did not use tubes and tunnels.

Just a few weeks ago J. Bloomquist, another of my graduate students, developed another test. If a housefly larvae is touched with a warm soldering iron, it curls around the end of the soldering iron. The needle is not hot; it is just warm and the very simple reflex response is consistent and repeatable. This kind of response can be produced in larval stages of other insects. I heard last week in Hamburg (at the 1984 Entomology Congress) that the response of the larvae of *Plutella xylostella*, the diamondback moth, is to back up in a rather dramatic fashion from a probe that is presented to its face.

Dr. Bloomquist treated 20 larvae with a dose of pyrethroid insecticide (fenvalerate), presented the stimulus and defined an ED_{50} as the dose producing a curling response in only 10 of the 20 larvae after 10 min. The data obtained which represents hundreds of larvae clearly shows a good difference between susceptible and *kdr*-maggots (Table 3). The ED_{50} for deltamethrin topically applied to these maggots is 0.113 µg. In the *kdr* strain the ED_{50} is 3.15 µg.

It is thought that the sodium channel of the *kdr*-strain is altered. We examined various drugs that act on the sodium channel to see how their response might be altered. The ED_{50} values for injected procaine were similar on both susceptible and *kdr* larvae, suggesting no change in the procaine site of action.

The same thing was true of tetrodotoxin; an ED_{50} of 89 vs. 116 is not statistically different, so that we suspect that the part of the sodium channel being af-

Table 3. Bioassay of sodium channel drugs on third instar larvae, *Musca domestica* by the thermal stimulus-response method. All values represent 240 maggots tested at 4 or more doses and replicated 3 times

Compound	Delivery	ED_{50} [a] (μg)	
		S_{NAIDM}	Kdr
Deltamethrin	Topical	0.113	3.15 [c]
Procaine	Inject	26.3	18.0
Tetrodotoxin	Inject	89.3 [b]	116.5 [b]
Scorpion venom	Inject	0.272	0.196
Aconitine	Inject	0.6	> 10.0 [c]

[a] Dose giving 50% paralysis 10 min after treatment
[b] Tetrodotoxin values for ED_{50} are in pg
[c] Significant differences between strains

fected by procaine or tetrodotoxin is not altered in this *kdr* strain. Aconitine in this test, and again you have to inject it because it is practically impossible to get a dose in from topical application, you do see a tremendous difference – similar to the deltamethrin difference 0.6 μg versus far greater than 10).

Components of scorpion venom that are suspected to act on certain parts of the sodium channel show a similar bioassay response, so what the *kdr* gene has done to the sodium channel alters responses to deltamethrin and aconitine but it does not affect responses to procaine, tetrodotoxin, and scorpion venom. With a little imagination, one can even think of designing an insecticide that acts on a portion of the sodium channel that is susceptible in these *kdr* strains.

4 Toxicokinetics

What concentration is needed for a given compound to be toxic? How much compound needs to penetrate an insect if you are designing an insecticide?

Figure 11 shows pharmacokinetic information from permethrin-treated adult houseflies. Doses applied in the nanogram range achieve concentrations in the hemolymph in the 10's of nanomolars. And 10^{-9} M is a relatively low concentration compared to other insecticides because the pyrethroids are fairly toxic.

For an insecticide like carbofuran the peak concentration that a toxic dose of carbofuran reaches is somewhere in the region of 10^{-5} M (Collins et al. 1983). Now for carbofuran to be toxic, it can be present at slightly less than that, but really we are dealing with the micromolar range or above in a practical sense. It is not necessary for a compound to be active in the nanomolar range to be insecticidal. So, when designing compounds or looking for activity, the micromolar range is a reasonable figure to consider.

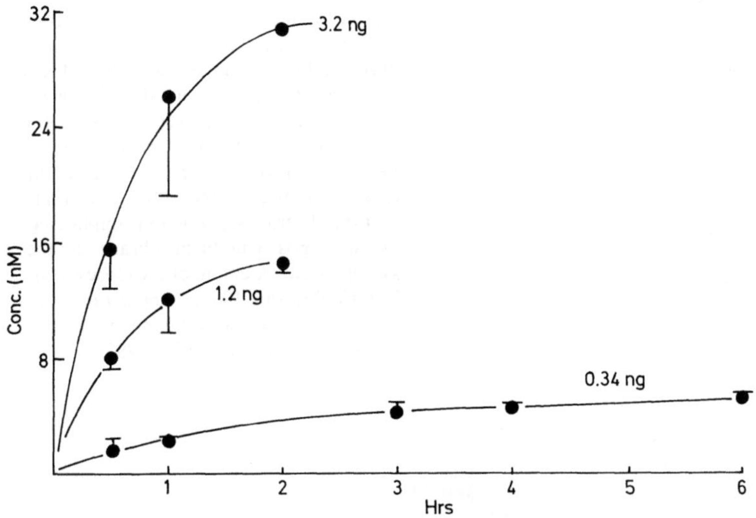

Fig. 11. Concentration of permethrin recovered from the hemolymph of adult female houseflies following topical treatment on the abdomen of the doses shown

5 Site of Action of Benzimidazoles

I have described above a series of bioassay tests and preparations that provide some insight into how insecticides might be acting and allow a fairly rapid assessment of questions, things like: Are they acting peripherally, are they acting centrally? I would like now to describe a series of compounds that was worked on by my graduate student, G. Bradfisch, the benzimidazoles.

Benzimidazoles have a reasonable amount of toxicity, 0.2 μg per fly, and these chemicals, as a matter of fact, are the result of serendipity. Dr. Kuwano, who works in the laboratory of Prof. Eto at Kyoto University, was looking for juvenile hormone analogs. You can recognize the farnesol chain here. But, he decided to hook the imidazole ring on one end of these compounds and instead of having just morphogenic effects, he found that they were neuroactive, because they produced nervous symptoms soon after treatment.

At the time of the *IUPAC* meeting in 1982, Dr. Kuwano and Dr. Eto asked if we would have a look at these compounds to find out how they were acting. This question became the basis for G. Bradfisch's Ph. D. Following the philosophy described above of looking at the muscle first, Dr. Bradfisch chose the Irving housefly larvae preparation and a fairly rapid decline in the neurally-evoked EPSP amplitude followed treatment by benzimidazole, particularly B-24. The spectacular thing about the responses was that they were not accompanied by hyperactivity; in the whole animal locomotor activity gradually and slowly declined following treatments. In fact, they were acting like rotenone, a general mitochondrial poison.

The first characteristic observed following treatment by benzimidazole was a decline in the postsynaptic potential in the muscle. Coinciding with this decline

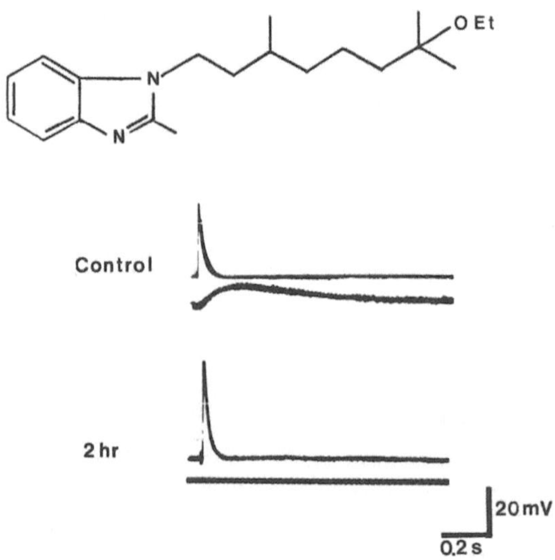

Fig. 12. Action of a benzimidazole derivative (B-24) 2 h after treatment with an LD_{95} dose. At this time the neurally-evoked excitatory postsynaptic potential (EPSP) was normal in appearance, even somewhat augmented, (*bottom* traces compared to untreated *top* traces); however, no tension increase accompanied the EPSP, indicating that the electrical-contraction coupling between the postsynaptic membrane and the sarcoplasmic reticulum of the muscle was lost. Calibration: 100 ms. *horizontal; vertical:* 20 mV EPSPs and 200 mg tension. (From Bradfisch and Miller 1985)

in EPSP amplitude, there was a loss of miniature postsynaptic potential activity. Long before the amplitude of EPSPs in the muscle of the maggot declined to zero, the muscle lost tension (Fig. 12). Two hours after treatment by B-24 the EPSP was potentiated, in other words, higher than normal at the time tension was lost. It was concluded from this that the benzimidazole compounds were acting both presynaptically and postsynaptically in some manner, and we suspected that they are acting somehow on a calcium channel (Bradfisch and Miller 1985).

The potential for finding out more about the site of action of benzimidazoles leading to synthesis of other more active molecules is wide open. The possibility of the calcium channel is just as exciting as the possibility for studying the GABA system (gamma aminobutyric acid) as sites of action leading to new neurotoxic compounds.

6 Picrotoxin

I have to end by mentioning that the benzimidazole compound was one of these serendipitous discoveries that occur from time to time, and yet benzimidazole chemistry has been around a long time. Another series of compounds, the picrotoxins, have been known as natural products. Picrotoxin is made up of equimolar amounts of picrotoxinin and the less active hydrated analog picrotin.

P. Usherwood used the picrotoxin mixture in 1965, as part of the proof for inhibitory innervation of insect muscles being GABA mediated (Usherwood and Grundfest 1965). As neurophysiologists, we have known about these compounds for years, and we have used them off the shelf. They selectively block the inhibitory synaptic transmission in insect muscles, they are active at 10^{-6} M, and they are reversible. I became interested in these because, first of all, I wondered why they were not particularly insecticidal and were not very active on the insect central nervous system. If one compares the structure of picrotoxinin (Fig. 13) with

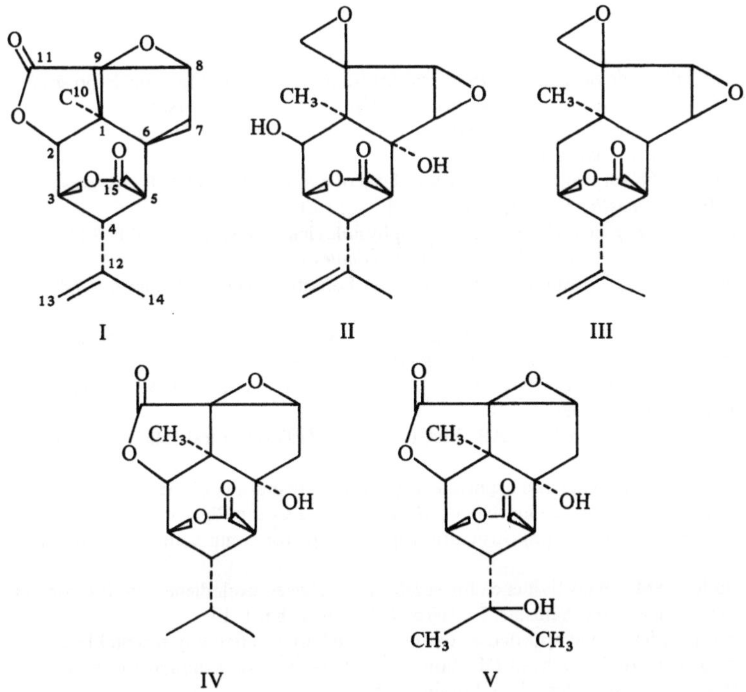

Fig. 13. Structures of picrotoxinin (*I*), tutin (*II*), coriamyrtin (*III*), α-dihydropicrotoxinin (*IV*) and picrotin (*V*)

the other members of the same family of compounds that were then known, you notice that the cyclohexane ring, which is the heart of the molecule, the propene group and the gamma lactone are all oriented in the same direction.

We did a structure-activity study in 1977, conducted by a student, M. Maynard, who at that time had just finished a Bachelor's Degree in chemistry. The study was supported entirely by our Dean at the time, Professor Mac Dugger. M. Maynard started by synthesizing a cyclohexane ring by using a modified Diels-Alder reaction. At the end of a year we found some convulsant activity in substituted γ-lactone cyclohexanes that was similar in potency to picrotin. It was suggested in our subsequent publication (Miller et al. 1979) that this sort of chemistry could directly lead to active compounds and then dropped it because of lack of support and because the problem at this point became one of concern to chemical industry. Rather than being forgotten, the picrotoxinin project led to some interesting observations. F. Matsumura has since suggested that there may be a similarity between the cyclohexane ring and insecticides like dieldrin or lindane (Matsumura and Ghiasuddin 1983). Well, it is funny what you wind up with when you start off in a given direction.

Mode of action studies can lead to possible candidate insecticides by studying natural products or following up on an accidental discovery. But more than that, mode of action studies give focus to structure activity studies. At least you know more precisely what activity you are looking at, and I believe that is a big help!

References

Adams ME (1978) Neurophysiological correlates of pyrethroid and DDT-type insecticide poisoning in the house fly *Musca domestica* L. Ph D Dissertation, Univ California, Riverside

Adams ME, Miller TA (1979) Site of action of pyrethroids; repetitive "backfiring" in flight motor units of house fly. Pestic Biochem Physiol 11:218–231

Adams ME, Miller TA (1980) Neural and behavioral correlates of pyrethroid and DDT-type poisoning in the house fly, *Musca domestica* L. Pestic Biochem Physiol 13:137–147

Bradfisch GA, Miller TA (1985) Behavioral and electrophysiological investigation of 1-(7-ethoxy)-geranyl-2-methyl benzimidazole on the house fly, *Musca domestica*. Pestic Sci (in press)

Busvine JR (1951) Mechanism of resistance to insecticide in houseflies. Nature (London) 168:193–195

Collins C, Kennedy JM, Miller TA (1979) Sublethal poisoning: A comparison of behavior and histological changes in house fly CNS. Pestic Biochem Physiol 11:135–158

Collins C, Kennedy JM, Miller TA (1983) Toxicokinetics of carbofuran in house fly, *Musca domestica* L. Pestic Biochem Physiol 20:25–38

Curtis DR, Davies J, Game CJA, Johnston GAR, McCulloch RM (1973) Central actions of shikimin and tutin. Brain Res 63:419–423

Irving SN, Miller TA (1980) Glutamate and aspartate as possible transmitters of the 'fast' and 'slow' neuromuscular junctions of body wall muscles of *Musca* larvae. J Comp Physiol 135:299–314

Jarboe CH, Porter LA, Buckler RT (1968) Structural aspects of picrotoxinin action. J Med Chem 11:719–731

Matsumura F, Ghiasuddin SM (1983) Evidence for similarities between cyclodiene type insecticides and picrotoxinin in the action mechanisms. J Environ Sci Health 18 B:1–14

Menn JJ (1983) Present insecticides and approaches to discovery of environmentally acceptable chemicals for pest management. In: Whitehead DL, Bowers WS (eds) Natural products for innovative pest management. Plenum Press, New York London, p 5

Miller TA, Maynard M, Kennedy JM (1979) Structure-activity studies of picrotoxinin. Pestic Biochem Physiol 10:128–136

Miller TA, Salgado VL, Irving SN (1983) The *kdr* factor in pyrethroid resistance. In: Georghiou GP, Saito T (eds) Pest resistance to pesticides: Challenges and prospects. Plenum Press, New York London, p 353

Mulloney B (1969) Interneurons in the central nervous system of flies and the start of flight. J Comp Physiol 64:243–253

Omer SM, Georghiou GP, Irving SN (1980) DDT/pyrethroid resistance interrelationships in *Anopheles stephensi*. Mosquito News 40:200–209

Salgado VL, Irving SN, Miller TA (1983) Depolarization of motor nerve terminals by pyrethroids in susceptible and *kdr*-resistant house flies. Pestic Biochem Physiol 20:100–114

Sparks TC, Shour MH, Wellemeyer EG (1982) Temperature-toxicity relationships of pyrethroids on three lepidopterans. J Econ Entomol 75:643–646

Usherwood PNR, Grundfest H (1965) Peripheral inhibition in skeletal muscle of insects. J Neurophysiol 28:497–518

Prospects of Exploitation of Insect Antijuvenile Hormones for Selective Insect Control

J. J. MENN [1]

1 Introduction

Much has been written on future approaches and strategies critical in the discovery process of novel, insect-specific insecticidal and insectostatic chemicals (Geissbühler 1981, Croft 1981, Menn and Henrick 1984). The fact that the Schering Corporation has dedicated this symposium to new approaches to "lead generation for insecticides" indicates the strength of the commitment that Schering, and currently several other major industrial concerns, have made to the field of crop, animal and human health protection.

We are looking primarily to chemical control of insect pests ten, twenty or more years into the future, that is, into the 21st century. Where will these chemical agents come from? Several important biochemical sites susceptible to interference by insecticide chemicals are discussed in Baillies, Millers, Fukamis and Eldefrawis chapters. Obviously, due to constraints of fund, time and manpower, research organizations will have to be very selective in the targets they choose to research and exploit. At Zoecon, we have targeted our recent research efforts at antijuvenile hormones (AJHs) as a promising approach to insect control.

2 Antijuvenile Hormones

From the early research on juvenile hormones (JHs) and juvenoids, it occurred to various investigators that disruption of JH activity at a critical stage in an insect's development would offer a superior approach to selective insect control (for reviews, see: Staal et al. 1981, Menn and Henrick 1981, Kramer and Staal 1981, Staal 1982).

2.1 Precocenes

Several model compounds with AJH activity resulting from different modes of action have been reported in the past several years. These include the chromenes, precocenes I and II (Fig. 1) (Bowers et al. 1976). These compounds were shown to be allatocidal, probably by forming highly reactive epoxides within the *corpora*

1 United States Department of Agriculture, Agricultural Research Service, National Program Staff, Beltsville, Maryland 20705, USA

Approaches to New Leads for Insecticides
Edited by von Keyserlingk et al.

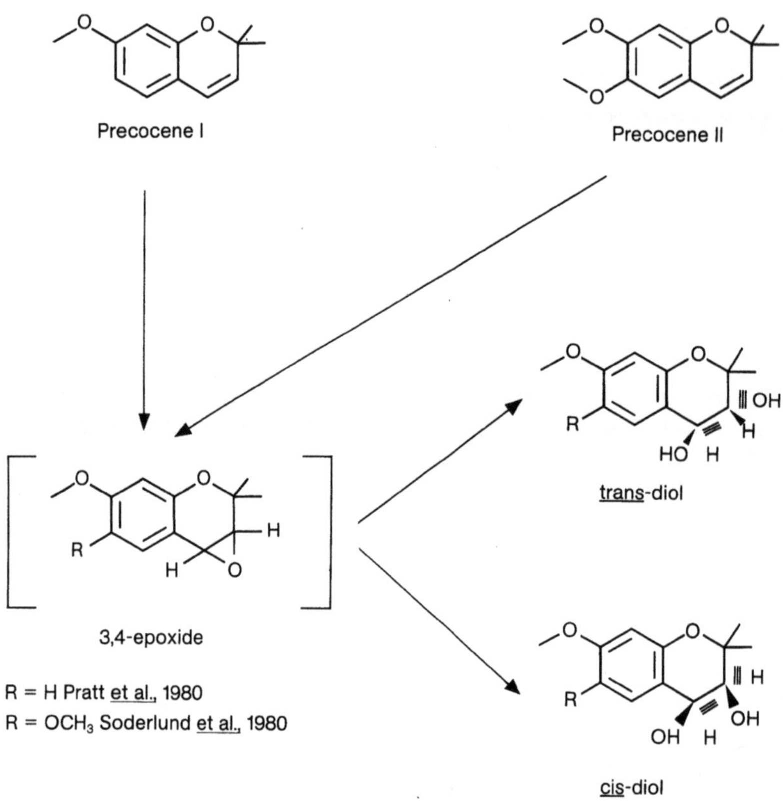

Fig. 1. Structures of *Precocenes I* and *II* and postulated bioactivity pathway in insects. (Menn and Henrick 1981)

allata (Fig. 1) (Söderlund et al. 1980, Brooks et al. 1979, Pratt et al. 1980). Most likely, the reactive epoxide alkylates sensitive site(s) within the *corpora allata*, causing their atrophy and the loss of JH secretion. From a practical point of view, the precocenes had limited utility, showing AJH activity in only a few insect species representative of the orders Hemiptera and Orthoptera.

2.2 Various Compounds Interfering with JH Biosynthesis

Very few compounds possessing AJH activity in lepidopterous insects have been discovered to date. Several of the older compounds with AJH activity include such diverse structures as ETB, piperonyl butoxide, EMD and FMev (Fig. 2). In larvae of the tobacco hornworm, *Manduca sexta*, these compounds have brought on morphogenetic symptoms in the form of black pigmentation, a syndrome characteristically associated with juvenile hormone deficiency (Staal 1982, Kramer et al. 1982 and 1983).

Quistad et al. (1981) have shown that fluoromevalonate (FMev) (Fig. 3) not only caused black pigmentation, but also induced precocious metamorphosis in

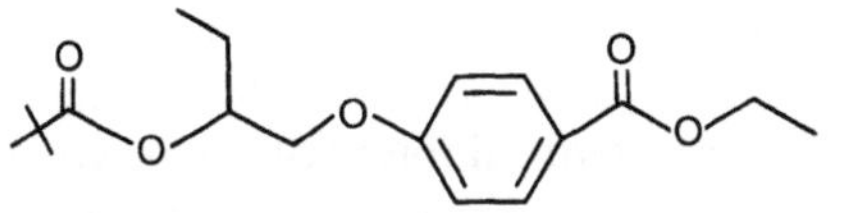

ETB, ethyl 4-[2-(tert-butylcarbonyloxy)-butoxy]benzoate
(Staal, 1977)

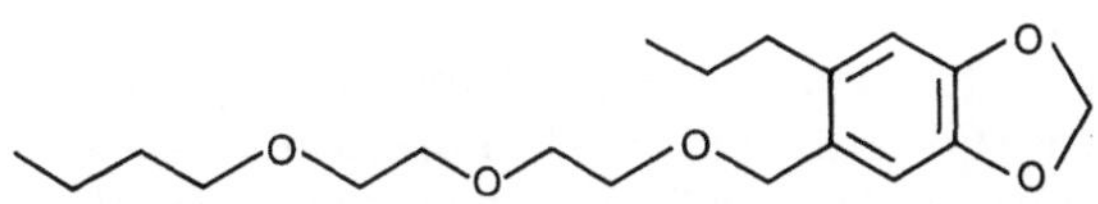

PB, piperonyl butoxide
(Staal et al., 1981)

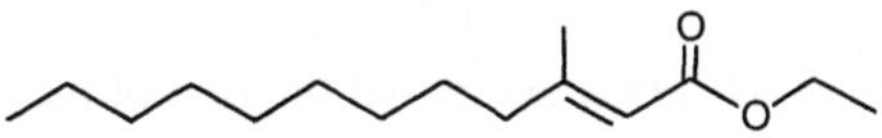

EMD, ethyl (E) -3-methyl-2-dodecenoate
(Staal et al., 1981)

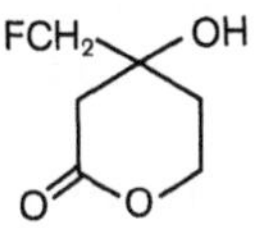

FMev, fluoromevalonolactone
(Quistad et al., 1981)

Fig. 2. Selected older compound with AJH activity

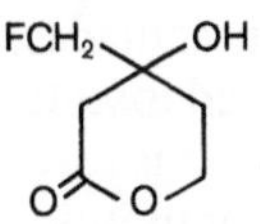

23
FMev

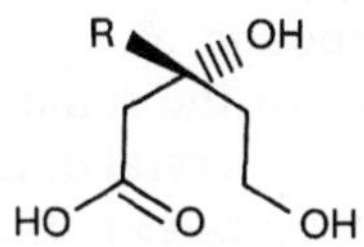

$R = CH_3$: Mev
$= C_2H_5$: HMev

Fig. 3. Structures of fluoromevalonolactone (*FMev*), mevalonate (*Mev*) and homoevalonate (*HMev*)

Table 1. Comparison of AJH activity of FMev on several lepidopterus species. (From Quistad et al. 1981)

Insect	Instar treated	$ED_{50\,mg\,g^{-1}}$ [a]
Manduca sexta (tobacco hornworm)	Third	0.7
Samia cynthia (cynthia moth)	Third	1.0
Phryganidia californica (California oak moth)	Third	1.0
Galleria mellonella (greater wax moth)	Fifth	40
Spodoptera exigua (beet armyworm)	Third	44
Heliothis virescens (tobacco budworm)	Third	>200 [b]

[a] The ED_{50} is expressed in milligrams of FMev per gram of insect
[b] The *H. virescens*, the ED_{50} was never achieved although a small percentage of the treated larvae did manifest premature pupation

larvae of several lepidopterous species (Table 1). FMev demonstrated significant AJH activity in *Samia cynthia, Phryganidia californica* and *M. sexta*. However, only weak activity was demonstrated against the important economic species, *Spodoptera exigua* and *Heliothis virescens*. No AJH activity was shown in the non-lepidopterous species: *Musca domestica, Aedes aegypti, Schistocerca nitens, Dermestes maculatus, Hypera postica* and *Oncopeltus fasciatus*. Synthetic efforts to optimize the AJH activity of FMev have only resulted in a number of compounds with AJH activity inferior to that of FMev (Quistad et al. 1982). Singer et al. (1959) have demonstrated that FMev was a potent inhibitor of the in vitro conversion of acetate and mevalonate to cholesterol by rat liver homogenates. Mode of action studies with FMev have shown that juvenile hormone titers in *M. sexta* hemolymph were significantly lowered after topical application of a dose causing precocious metamorphosis (Edwards et al. 1983). Furthermore, concurrent treatment of FMev with a juvenoid "rescued" the insect from AJH effects (Quistad et al. 1981). F.C. Baker (1979 unpublished data) also found that FMev incubated in homogenates of *corpora allata* of *M. sexta* decreased the rate of metabolism of [^{14}C]-mevalonate and [^{3}H]-homomevalonate to subsequent isoprenoid precursors. The latter steps are shown in the schematic outline of the biosynthetic pathway of insect juvenile hormones from acetyl CoA plus propionyl CoA (Schooley and Baker 1985) (Fig. 4).

Table 2 compares the bioevaluation data for FMev and a newer AJH compound (Compound X), synthesized in the Zoecon Research Laboratories, to that of conventional insecticides.

It should be noted that, here, we are comparing LD_{50} values for conventional insecticides with ED_{50} values for FMev and Compound X.

While FMev still compares poorly in toxicity to the standard insecticides, Compound X compares reasonably well in activity to several of the listed insecticides, especially by comparing activity on *Spodoptera exigua*. Based on this lead, we are continuing our synthetic efforts to discover even more potent compounds with AJH activity.

Relatively few new reports on compounds with significant AJH activity have appeared in the literature. Several of the more prominent chemical prototypes are shown in Figs. 2 and 5. The compound ETB (Fig. 2) shows selective AJH activity in *Manduca sexta* (Staal 1977). At low doses in the diet (1 ppm), it produced black

Table 2. Comparison of potency of two synthetic antijuvenile hormone agents with standard insecticides[a]

	Topical LD_{50} µg third instar^{-1}		
	Heliothis virescens	*Spodoptera exigua*	*Manduca sexta*
Fluoromevalonate (FMev)	100[b]	40[b]	16[b]
Compound X	6.9[b]	1.9[b]	0.51[b]
DDT	36	2.3	>100
Malathion	4.8	0.81	35
Methyl parathion	0.49	0.059	1.3
Acephate	1.3	4.3	30
Carbaryl	0.58	1.01	25
Methomyl	0.11	0.15	6.7
Carbofuran	0.77	1.7	0.45
Permethrin	0.018	0.015	0.029

[a] Data from D. Cerf, Zoecon Corporation
[b] ED_{50} values

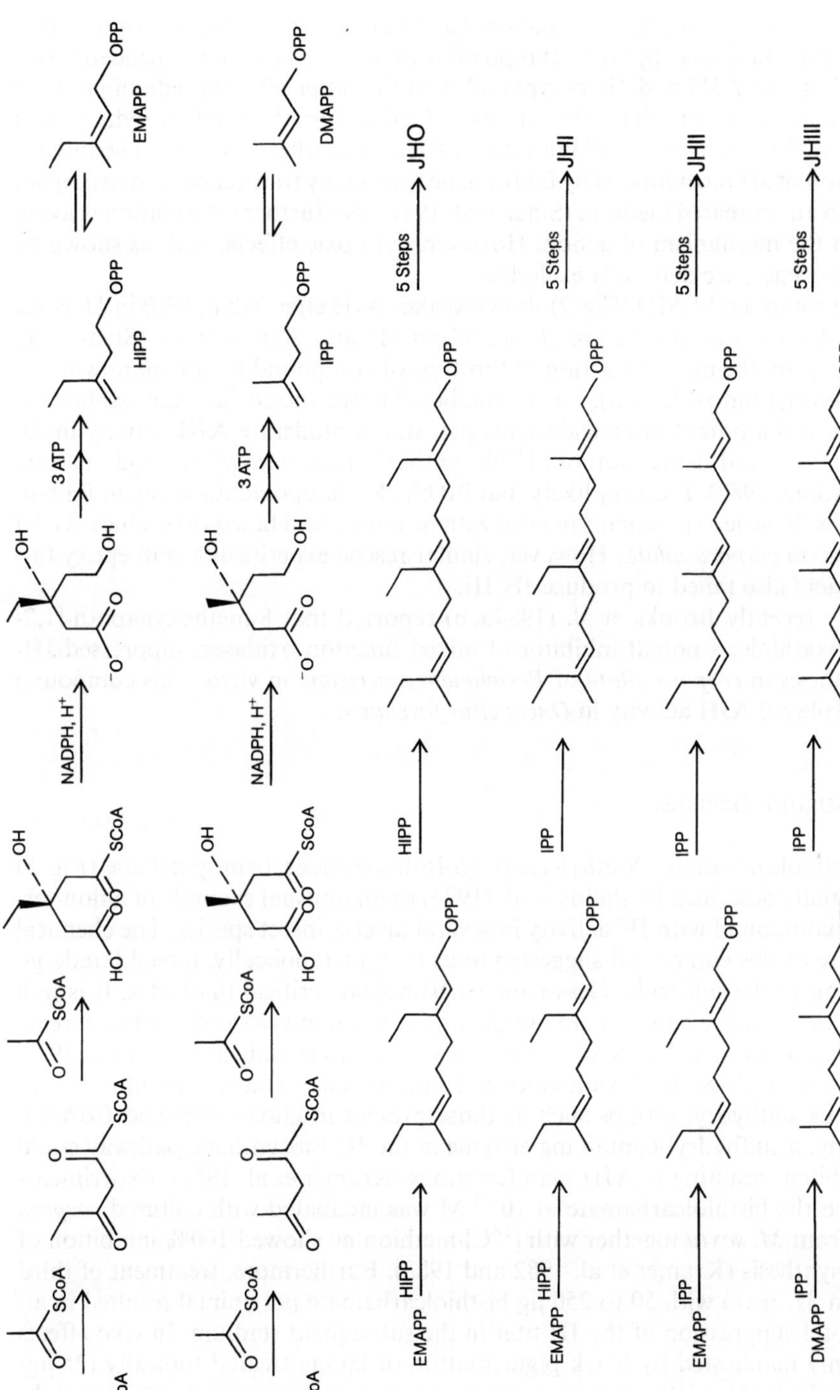

Fig. 4. Biosynthetic scheme of juvenile hormones from acetyl or propionyl – CoA. Abbreviations: *HIPP* homoisopentenyl pyrophosphate; *IPP* isopentenyl pyrophosphate; *EMAPP* ethylmethylallyl pyrophosphate; *DMAPP* dimethylallyl pyrophosphate (Schooley and Baker 1985, Fig. 9)

pigmentation and precocious pupation. However, at higher doses (3 ppm) no premature pupation was apparent. It is possible that it acts in a dose-dependent manner as both an AJH and JH compound with the latter effects predominating at higher concentrations. Its AJH activity is further corroborated by reduction of JH I and II titers in vivo and inhibition of JH biosynthesis in vitro. The in vitro inhibition of JH biosynthesis by ETB can be rescued by transfer of treated *corpora allata* to an untreated medium (Staal et al. 1981). No further information is available on the mechanism of action. However, cytotoxic effects, such as shown by the precocenes, are definitely excluded.

The compound EMD (Fig. 2) shows weaker AJH effects than ETB in *M. sexta* and *Heliothis virescens*. It too shows mixed JH and AJH activity (Staal et al. 1981). Again, the mode of action of this type of compound is still unknown.

Piperonyl butoxide (PB), one of the best known mixed function oxidase inhibitors and a potent insecticide synergist, shows moderate AJH activity in *M. sexta* in vivo and depression of JH biosynthesis in *corpora allata* organ culture assays (Staal 1982). It is very likely that PB blocks the epoxidation step in JH biosynthesis. Rescue experiments in vitro with farnesoic acid failed to produce JH III in *M. sexta corpora allata*. However, similar rescue experiments with epoxy farnesoic acid also failed to produce JH III.

Very recently Brooks et al. (1984a, b) reported that 8-methoxynaphth-[1,2-d]-1,3-oxathiole, a potent inhibitor of mixed function oxidases, suppressed JH-biosynthesis in *corpora allata* of *Periplaneta americana* in vitro. This compound also displayed AJH activity in *Oncopeltus fasciatus*.

2.3 Bisthiolcarbamates

The bisthiolcarbamate, *N*-ethyl-1,2-bis(*S*-isobutylthiocarbamoyl)ethane (Fig. 5) was initially described by Pallos et al. (1976) as an unusual example of a non-terpenoid compound with JH activity in several diverse insect species. The chemical structure of this compound suggested to us that, metabolically, it could undergo oxidation to the sulfoxide, cleave and carbamoylate critical thiol sites. It is well known that thiolcarbamates are rapidly metabolized in vivo and in vitro to reactive sulfoxides and possibly very short-lived sulfones (Hubbell and Casida 1977, DeBaun et al. 1978). Both sulfoxides and sulfones may readily carbamoylate endogenous sulfhydryl groups such as those present in glutathione and CoASH. Similarly, a sulfhydryl-containing enzyme in the JH biosynthetic pathway could be inhibited, resulting in AJH manifestations (Kramer et al. 1982). Experiments in which the bisthiolcarbamate at 10^{-4} M was incubated with cultured *corpora allata* from *M. sexta* together with [^{14}C]-methionine showed 100% inhibition of JH biosynthesis (Kramer et al. 1982 and 1983). Furthermore, treatment of third stadium *M. sexta* with 50 to 250 μg bisthiolcarbamate per animal resulted in almost total suppression of the JH titer in the subsequent stadium. In vivo effects were only manifested by black pigmentation of larvae treated topically (25 μg/larva) or by feeding (10 ppm); no premature pupation or prepupal cuticle was observed following treatment with this compound. However, at higher doses typical JH effects (paleness) were also observed (Kramer et al. 1983).

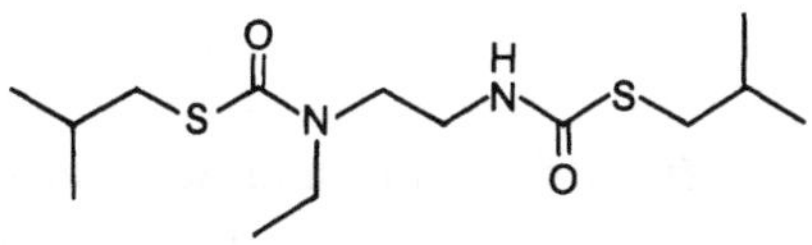

Bisthiolcarbamate, N-ethyl-2-bis(S-isobutyl-thiocarbamoyl)ethane (Kramer et al., 1982)

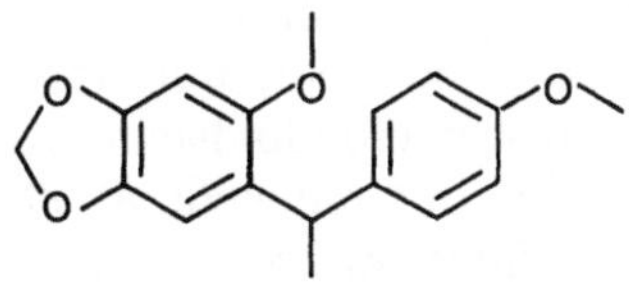

J-2710, 5-methoxy-6-[1-(4-methoxyphenyl)ethyl]-1,3-benzodioxole (Van Mellaert et al., 1983)

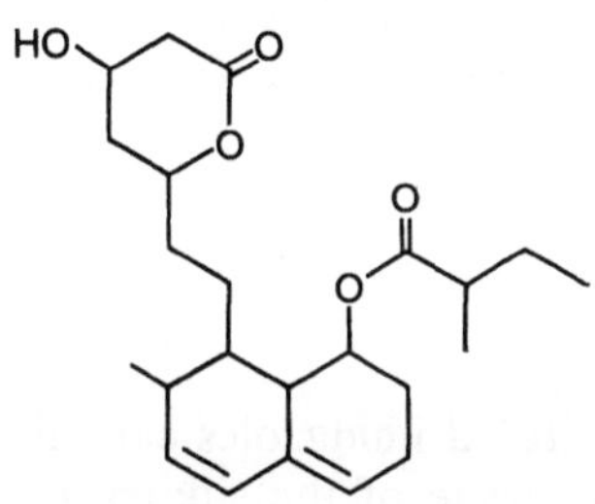

Compactin, ML-236B (lactone form) (Hiruma et al., 1983)

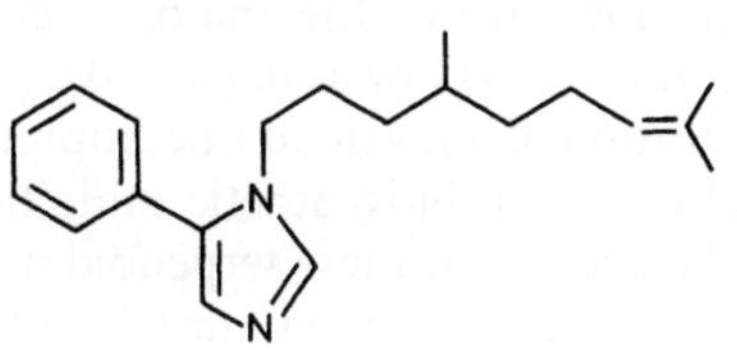

Terpenoid imidazole, 1-citronellyl-5-phenylimidazole (Kuwano et al., 1983)

Fig. 5. Structures of several newer compounds with AJH activity

The weak activity of the bisthiolcarbamate may have resulted from rapid degradation of the compound in vivo. Extensive studies in several laboratories have shown that the same metabolic processes which "activate" the parent thiolcarbamates are also essential in their degradation (Menn 1978, Schuphan et al. 1981). Furthermore, we know virtually nothing concerning the uptake, transport and action of xenobiotics in the *corpus allatum*. It is not clear yet whether this compound exerts its action via JH biosynthesis inhibition or JH agonist feedback inhibition via the insect brain (Kramer et al. 1983). Nevertheless, this may be an interesting model for possible additional synthesis of more active and stable analogs.

2.4 Benzodioxoles

Another type of compound with AJH activity was recently reported by Van Mellaert et al. (1983). These authors reported that benzyl-1,3-benzodioxoles and benzylphenols, which were previously shown to be insect chemosterilants (Jurd et al. 1979), also displayed AJH activity in the *Galleria mellonella* wax test. Compound

J-2710 (Fig. 5) reduced the activity of JH 111 100,000-fold when assayed at 1% concentration in the *Galleria* bioassay. These authors theorized that the known chemosterilant action of these compounds in *Musca domestica* and *Sarcophaga bullata* is associated with AJH effects. They proposed that, possibly, benzodioxoles are in competition with JH for the JH receptors. Blocking of the JH receptor would result in disruption of events leading to egg maturation.

Compactin (ML-236 B) (Fig. 5) is a hypocholesterolemic agent; more specifically, it inhibits microsomal HMG-CoA reductase (Brown and Goldstein 1980). Hiruma and coworkers reported that repeated injection of Compactin at 50 μg into penultimate larvae of the cabbage armyworm, *Mamestra brassicae* caused both black pigmentation and precocious metamorphosis in the last larval instar. The low level of AJH activity (Table 3) contrasts with its remarkable in vitro activity in the rat liver system. Since only multiple doses were effective, Compactin appears to be degraded rapidly in the living insect.

2.5 Imidazoles

Kuwano et al. (1983) have shown that several 4-substituted imidazoles caused precocious metamorphosis in third and fourth instar larvae of the silkworm, *Bombyx mori,* after topical application at very low doses. One analog, 1-citronellyl-5-phenylimidazole (Fig. 5) was effective at doses as low as 0.5 to 1.0 μg per larva. The AJH effect was blocked by administration of a juvenoid (methoprene). It is possible that compounds of this type also act via biosynthetic inhibition. However, no direct evidence exists to date. The activity of these terpenoid imidazoles appears to be highly selective in insects. Unfortunately, in our laboratory the remarkable AJH activity in *B. mori* did not extend to several diverse insect species that we tested (Table 3).

Table 3. Comparison of potency of several compounds with AJH activity[a]

	ED_{50} topical application (μg 3rd instar larva^{-1})		
	Heliothis virescens	*Spodoptera exigua*	*Manduca sexta*
FMev	100	40	16
Compound X	6.9	1.9	0.51
Compactin (ML-236B)	>100	>100	>100
Terpenoid imidazole	>250	>250	>250
EMD	>100	–	>100
ETB	>250	–	>250
PB	>100	–	180
Bisthiolcarbamate	>250	–	>250
J-2710	>100	–	>250

[a] Data from D. Cerf, Zoecon Corporation
[b] ED_{50} = Effective dose causing premature metamorphisis in 50% of test population

2.6 Outlook

In this review, I have briefly examined ten chemically diverse AJH active compounds. However, none of these compounds show sufficient activity for practical application (Table 3). Nevertheless, Compound X approaches most closely what may be termed as a chemical lead for further synthesis optimization.

Presently, Zoecon and several other companies and institutions are continuing with synthetic efforts in this exciting endeavor which eventually may pave the way toward the development of exquisitely selective and potent insect control chemicals.

References

Bowers WS, Ohta T, Cleere JS (1976) Discovery of insect antijuvenile hormones in plants. Science 193:542–547

Brooks GT, Pratt GE, Jennings RC (1979) Action of precocenes in milkweed bugs (*Oncopeltus fasciatus*) and locusts (*Locusta migratoria*). Nature (London) 281:570–572

Brooks GT, Pratt GE, Ottridge AP, Mace DW (1984a) The inhibition of JH III biosynthesis in *corpora allata* of the cockroach *Periplaneta americana* in vitro. Proc 1984 Br Crop Protect Conf 3:881–886

Brooks GT, Pratt GE, Ottridge AP, Cocks JA (1984b) Effects of inhibitors on JH III biosynthesis in *corpora allata* of the cockroach *Periplaneta americana* L. in vitro. J Pestic Sci (Tokyo) 9:211–214

Brown MS, Goldstein JL (1980) Multivalent feedback regulation of HMG-CoA reductase, a control mechanism coordinating isoprenoid synthesis and cell growth. J Lipid Res 21:505–517

Croft BA (1981) Use of crop protection chemicals for integrated pest control. Philos Trans R Soc London Ser B 295:125–140

DeBaun JR, Bova DL, Tseng CK, Menn JJ (1978) Metabolism of [ring-^{14}C] Ordram (molinate) in the rat. J Agric Food Chem 26 (5):1098–1104

Edwards JP, Bergot BJ, Staal GB (1983) Effects of three compounds with antijuvenile hormone activity and a juvenile hormone analogue on endogenous juvenile hormone levels in the tobacco hornworm, *Manduca sexta*. J Insect Physiol 29:83–89

Geissbühler H (1981) The agrochemical industry's approach to integrated pest control. Philos Trans R Soc London Ser B 295:111–123

Hiruma K, Yagi S, Endo A (1983) ML-236B (Compactin) as an inhibitor of juvenile hormone biosynthesis. Appl Entomol Zool 18 (1):111–115

Hubbell JP, Casida JE (1977) Metabolic fate of the N,N-dialkylcarbamoyl moiety of thiocarbamate herbicides in rats and corn. J Agric Food Chem 25 (2):404–413

Jurd L, Fye RL, Morgan J Jr (1979) New types of insect chemosterilants benzylphenol and benzyl-1,3-benzodioxole derivatives as additives to housefly diet. J Agric Food Chem 27 (5):1007–1016

Kramer SJ, Staal GB (1981) In vitro studies on the mechanism of action of antijuvenile hormone agents in larvae of *Manduca sexta*. In: Pratt GE, Brooks GT (eds) Juvenile hormone biochemistry. Elsevier/North-Holland Biomed Press, Amsterdam, New York

Kramer SJ, Tsai LW, Lee SF, Menn JJ (1982) Inhibition of juvenile hormone biosynthesis in the tobacco hornworm, *Manduca sexta*, by a bisthiolcarbamate juvenoid and related compounds. Pestic Biochem Physiol 17:134–141

Kramer SJ, Baker FC, Miller CA, Cerf DC, Schooley CA, Menn JJ (1983) Juvenile hormone antagonists. In: Miyamoto J et al. (eds) Human welfare and the environment. Pergamon Press, Oxford, New York

Kuwano E, Takeya R, Eto M (1983) Terpenoid imidazoles: New anti-juvenile hormones. Agric Biol Chem 47 (4):921–923

Mellaert H Van, DeLoof A, Jurd L (1983) Anti-juvenile hormone effects of newly described chemosterilants: Benzyl-1,3-benzodioxoles and benzylphenols. Entomol Exp Appl 33:83–88

Menn JJ (1978) Comparative aspects of pesticide metabolism in plants and animals. Environ Health Perspect 27:113–124

Menn JJ, Henrick CA (1981) Rational and biorational design of pesticides. Philos Trans R Soc London Ser B 295:57–71

Menn JJ, Henrick CA (1984) Newer chemicals for insect control. In: Chemicals for plant and animal protection. Beltsville Symp Agric Res, vol VIII. Agricultural chemicals of the future, May 16–19, 1983, Beltsville, Maryland

Pallos FM, Letchworth PE, Menn JJ (1976) Novel nonterpenoid insect growth regulators. J Agric Food Chem 24:218–221

Pratt GE, Jennings RC, Hamnett AF, Brooks GT (1980) Lethal metabolism of precocene I to a reactive epoxide by locust *corpora allata*. Nature (London) 284:320–323

Quistad GB, Cerf DC, Schooley DA, Staal GB (1981) Fluoromevalonate acts as an inhibitor of insect juvenile hormone biosynthesis. Nature (London) 289:176–177

Quistad GB, Staiger LE, Cerf DC (1982) Preparation and biological activity of potential inhibitors of insect juvenile hormone biosynthesis. J Agric Food Chem 30:1151–1154

Schooley DA, Baker FC (1985) Juvenile hormone biosynthesis. In: Kerkut/Gilbert (eds) Comprehensive insect physiology biochemistry and pharmacology, vol VII. Pergamon Press, Oxford, New York, in press

Schuphan I, Segall Y, Rosen JD, Casida JE (1981) Toxicological significance of oxidation and rearrangement reactions of S-chloroallyl thio- and dithiocarbamate herbicides. In: Rosen JD, Magee PS, Casida JE (eds) Sulfur in pesticide action and metabolism. ACS Symp Ser 158:65–82

Singer FM, Januszka JP, Borman A (1959) New inhibitors of in vitro conversion of acetate and mevalonate to cholesterol. Proc Soc Exp Biol Med 102:370–373

Söderlund DM, Messequer A, Bowers WS (1980) Precocene II metabolism in insects: Synthesis of potential metabolites and identification of initial in vitro biotransformation products. J Agric Food Chem 28 (4):724–731

Staal GB (1977) Insect control with insect growth regulators based on insect hormones. Pontif Accad Sci Scr Varia 41:353–383

Staal GB (1982) Insect control with growth regulators interfering with the endocrine system. Entomol Exp Appl 31:15–23

Staal GB, Henrick CA, Bergot BJ, Cerf DC, Edwards JP, Kramer SJ (1981) Relationships and interactions between JH and anti JH analogues in Lepidoptera. In: Sehnal F, Zabza, Menn JJ, Cymborowski B (eds) Regulation of insect development and behaviour. Wroclaw Tech Univ Pres, Wroclaw, Poland, pp 323–340

Insecticides as Inhibitors of Respiration

J.-I. Fukami[1]

1 Introduction

Under aerobic conditions insects meet their energy requirements by respiration. This is essentially the oxidation of biological fuel molecules. Energy is liberated via the respiratory chain and stored as energy-rich phosphate during the process of oxidative phosphorylation. The many chemical steps involved in this process of respiration and in the subsequent conservation of the derived energy in the form of ATP are catalyzed by a number of enzymes. Here we find differences between insects and mammals.

The metabolic costs of insect flight are tremendous. The oxygen consumption at rest is 10 $\mu l\ g^{-1}\ min^{-1}$. During flight it increases 15–20 times to 166–500. Sacktor (1965) has reported that in insects the oxygen consumption increases over 100-fold during the transience from rest to flight.

In this report I will focus on the differences in respiration between insects and mammals and the interactions between insecticides and the respiratory system in insects.

2 Biochemistry of the Respiratory System

Insect muscles can be divided into two groups. The first, characterized by jumping muscles, contains a low α-glycerophosphate dehydrogenase activity and high lactate-dehydrogenase activity. The second group, which is characterized by insect flight muscles, contains low lactic dehydrogenase and high α-glycerophosphate dehydrogenase. Under anaerobic conditions tissues of the jumping muscle group produce lactate whereas tissues of the flight muscle group produce roughly equimolar amounts of a α-glycerophosphate and pyruvate.

The leg muscle of the locust has 33 and 117 units of α-glycerophosphate and lactate dehydrogenase, respepctively, whereas the flight muscle of the locust has 167 units of α-glycerophosphate dehydrogenase but only two units of lactate dehydrogenase. In vertebrates the lactate dehydrogenase of the sceletal muscle is very high. This tendency in the rat sceletal muscle differs sharply from that in insect flight muscles (Sacktor 1965).

1 Laboratory of Insect Toxicology, Institute of Physical and Chemical Research, Wako-shi, Saitama, 351-011, Japan

Approaches to New Leads for Insecticides
Edited by von Keyserlingk et al.

The glycolysis and the α-glycerophosphate shuttle during insect flight has been investigated. The α-glycerophosphate which is produced by the insect flight muscle cell may originate either from the catabolism of phospholipids or from dihydroxyacetone phosphate (DHAP) derived from the classical glycolytic pathway of Embden-Meyerhof. In vertebrates, α-glycerophosphate is not formed since the NADH formed during the glycolytic oxidation of 3-phospho-glyceroaldehyde is rapidly re-oxidized to NAD during the subsequent conversion of pyruvate to lactate by lactate dehydrogenase (Estabrock and Sacktor 1958, Chance and Sacktor 1958, Marquardt and Brosemer 1966).

In vertebrates the oxidation of α-glycerophosphate by mitochondrial α-glycerophosphate dehydrogenase (αGPII) is low in liver mitochondria but occurs to a limited extent in muscle and brain mitochondria. In the flight muscle of the insect, α-glycerophosphate is the substrate for which the mitochondria have the highest activity.

These observations and the fact that α-glycerophosphate dehydrogenase activity can exceed the rate of succinate oxidation by a factor over 10 suggest that, on a kinetic basis, the TCA cycle may not play a significant role in the insect flight muscle metabolism (Chance and Sacktor 1958) (see Fig. 1).

The respiratory chain is commonly known as the mitochondrial electron transport system. It receives electrons from the TCA cycle, the fatty acid cycle, or NAD- or flavin-linked enzyme systems and transfers them through a series of redox intermediates to oxygen. The major oxidation-reduction components that occur in the respiratory chain in mitochondria are (a) NAD or NADP dehydrogenases (Pyridine nucleotide linked dehydrogenases) which require either NAD or NADP as coenzyme, (b) flavin-linked enzymes which contain flavin adenine dinucleotides (FAD) or flavin mononucleotides (FMN) as their prostetic groups, (c) the cytochromes, which contain an iron-phorsphyrin ring system, and (d) the lipid soluble quinones (Fukami 1976) (see Fig. 2).

The energy sources utilized by flying insects are of two major types. In general, the insects such as the locust and the butterfly which can fly for extended periods of time utilize fat as the major source of energy during flight. In contrast, insects such as the housefly and the bee which usually fly for much shorter periods utilize more carbohydrates as their main source of flight energy (Sacktor 1965). The significance of the oxidation of α-glycerophosphate and pyruvate in energy production during insect flight has already been discussed. However, isolated mitochondria of the blowfly rapidly lose their capacity to oxidize pyruvate. In the blowfly this loss can be reversed by proline. But in the rat it can be reversed by TCA cycle intermediates or glutamate. These findings, combined with the rapid oxidation of proline in insects suggest that, in vivo, flight muscle mitochondria may be deficient in the TCA cycle intermediates and that these are generated from proline. Proline is converted to L-glutamate as follows by a two-step oxidative reaction through the intermediate Δ-Pyrroline-5-carboxylate. The L-glutamate thus formed can generate TCA cycle intermediates by further mitochondrial oxidation (Sacktor and Childress 1967) (see Fig. 3). The cytochrome pigment of the meso- and metathoracic muscle of the cockroach can be separated following its reduction with sodium hydrosulfite. The α bands of the cytochrome occur at the usual position (a, 603 nm; b, 562 nm; c, 551 nm). In the Soret region the absorption

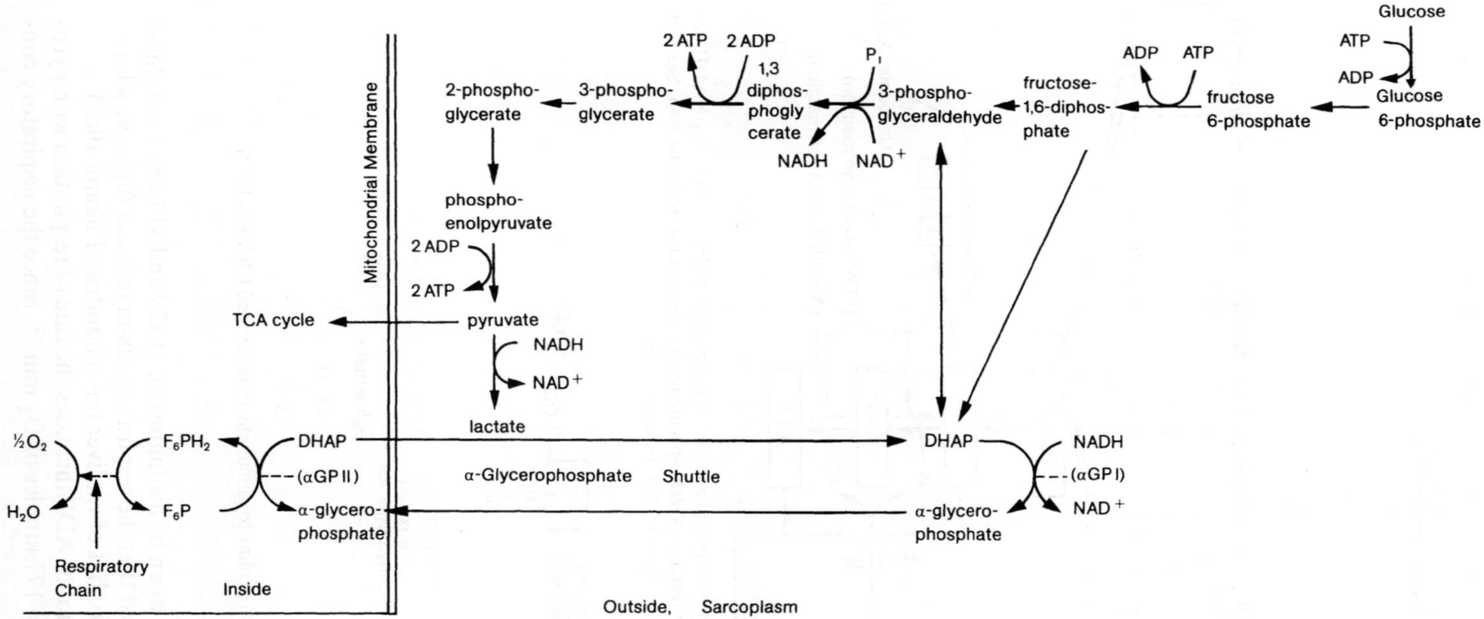

Fig. 1. Glycolysis and the α-glycerophosphate shuttle in insect flight muscle. Abbreviations: *αGPII* α-glycerophosphate dehydrogenase in mitochondria; F_GP mitochondrial flavoprotein; *DHAP* dihydroxyacetone phosphate; *α-GPI* NAD-linked α-glycerophosphate dehydrogenase in sarcoplasm. (Fukami 1976, from Chance and Sacktor 1958)

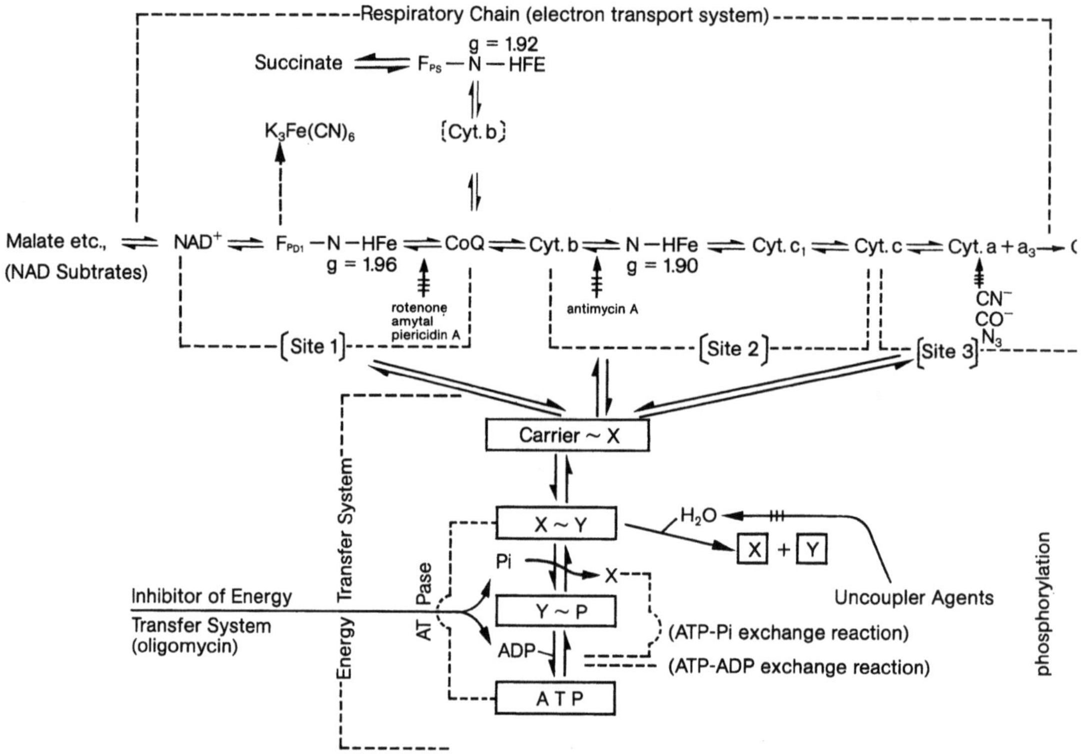

Fig. 2. Respiratory enzyme system of mammalian liver mitochondria. Abbreviations: F_{PD} NADH dehydrogenase; F_P succinate dehydrogenase; *N-HF$_e$* specific protein containing nonheme iron. See text for explanation. (From Lardy and Ferguson 1969, Fukami 1976)

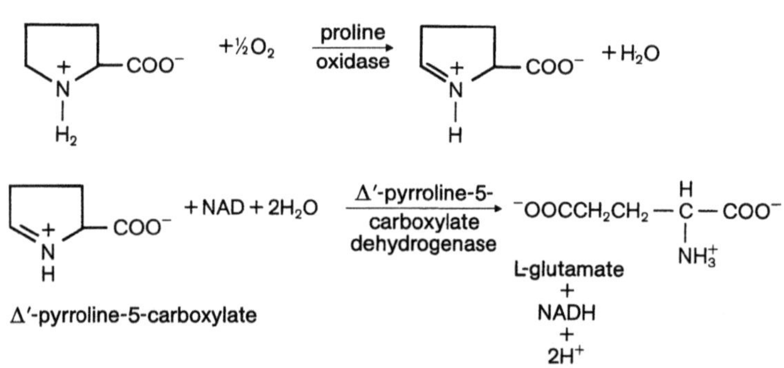

Fig. 3. The metabolic pathway from proline to glutamate. (Sacktor and Childress 1967)

bands of cytochrome b and c can be distinguished at 429 and 419 nm. From these studies, it has been concluded that the respiratory chain in insect flight muscle mitochondria is very similar to that of rat liver mitochondria (Fukami 1961).

The addition of 0.64 μmol of ADP increases the rate of respiration with pyruvate as substrate from 40 to 177 μmoles of O_2 min^{-1}. Since the respiratory con-

Table 1. Respiratory control and ADP: oxygen with several substrates. (Matsuda and Fukami 1972)

	American cockroach muscle				Rat liver			
	Respiratory activity		R.C.	ADP:O	Respiratory activity		R.C.	ADP:O
	State 3	State 4			State 3	State 4		
Pyruvate	227	46	4.9	2.9	15	13	1.2	1.8
Citrate	8	8	1.0		5	5	1.0	
α-Ketoglutarate	77	77	1.0		19	19	1.3	1.4
Succinate	195	195	1.0		69	27	2.6	1.9
Fumarate	21	21	1.0					
Malate	19	19	1.0					
Glutamate	35	35	1.0		18	18	1.0	
α-Glycerophosphate	215	215	1.0		13	13	1.0	

trol is defined as the ratio of the respiration in the presence of an acceptor (ADP) (state 3) to that after added ADP has been converted to ATP (state 4), the ratio in the example is 177/40 or 4.4. In general, mitochondria with high respiratory control ratios also exhibit high P:O ratios (Chance and Williams 1956, Matsuda and Fukami 1972).

Comparative studies have established that oxidative phosphorylation in isolated insect muscle mitochondria is usually less efficient than in those from vertebrate liver or heart.

The results show that although respiratory activity in American cockroach muscle mitochondria is high a respiratory control ratio greater than 1.0 was obtained only with pyruvate out of a series of different substrates.

In similar experiments with rat liver mitochondria, R. C. values greater than 1.0 were obtained with pyruvate, succinate and α-ketoglutarate (Matsuda and Fukami 1972, Ilivicky et al. 1967) (see Table 1).

Adenosine nucleotide contents of insect and vertebrate tissues have been investigated. The high content of ATP in both insect and rat muscle as well as in the tissue of developing adults of *Hyalophora* is probably related to their high level of oxidative metabolism. Thus, ATP/ADP ratios would suggest little or no impairment of glycolysis in these tissues. By contrast, the low ATP content for liver and the housefly muscle in the early phase of adult development may reflect either impaired ATP production or a high ATP utilization for biosynthetic activity.

3 Interaction of Insecticides with the Respiratory System

In general, inhibitors of the respiratory enzyme systems can be separated into three types: (a) inhibitors of the respiratory chain (the electron transport system) (type I), (b) inhibitors of the oxidative phosphorylation (type II), and (c) inhibitors of the energy transfer system (type III).

Type I inhibitors of the respiratory chain may be divided into four major groups according to the site at which they exert their effects. These are (a) inhibitors acting between NAD and CO-enzyme Q, (b) the inhibitors of succinate oxidation, (c) inhibitors acting between cytochrome b and c_1, and (d) cytochrome c oxidase inhibitors.

The site at which inhibition occurs can be studied with the oxygen electrode. These inhibitors completely block mitochondrial respiration, and since oxidative phosphorylation is coupled with the respiratory chain, ATP is not produced in their presence. This can be demonstrated for instance by the effect of antimycin A on succinate oxidation before (a) and after (b) the addition of ADP. It can clearly be seen that antimycin A causes an immediate and complete block of respiration in both cases.

3.1 Inhibitors of the Respiratory Chain

3.1.1 Inhibitors Acting Between NAD and Coenzyme Q

Inhibitors blocking the respiratory chain specifically between NAD and Coenzyme Q are the insecticide rotenone and the antibiotic piericidin A (Fukami 1961, Jeng et al 1968).

Rotenone and other rotenoids constitute most of the insecticidally active constituents of certain plants, rotenone being the most important.

Rotenone is used as an insect poison for the control of pests on plants and animals, as a fish poison to control populations in reservoirs, lakes and streams, and to eliminate undesirable species. As an insecticidal chemical, it has good residual activity and does not induce residue tolerances when applied to growing crops in accordance to good agricultural practice.

Rotenone depresses respiration and heart activity in insects. Paralysis occurs from the initial stages of intoxication and these symptoms differ greatly from those caused by pyrethrins, nicotine, DDT and organophosphorus insecticides. Although the primary biochemical lesion causing these effects was not known at that time. Tischler (1936) proposed that the primary action of rotenone may be due to the blockade of oxygen uptake in insects. From this suggestion, it has long been assumed that the primary action of rotenone is an inhibitory one on the respiratory enzymes in insects and that such inhibition causes a disturbance in the physiological function of one or more organs. It has been demonstrated that the cockroach was affected with paralysis and that its respiration fell slowly after rotenone injection (Fukami 1954).

Electrophysiological experiments with nerves and muscles from poisoned insects revealed an initial decrease in spontaneous discharges in the central nervous system followed by a conduction block. The ability of the muscle to contract was progressively blocked after the rotenone treatment. Failure to observe any indication of enhanced nervous activity in the early stage of rotenone poisoning suggested little or no direct action on the nervous system and it was assumed that the primary action of rotenone was the inhibition of cell respiration in the nervous and muscle tissues. The degree of TTC (2,3,5-triphenyl tetrazolium chloride) staining of nerve and muscle tissues was depressed in the paralyzed roach treated with rotenone or methyl parathion, but was not depressed with DDT or pyrethrins. During rotenone intoxication of the American cockroach the depression of staining was recognized just after knockdown (Fukami 1956). The change in oxygen uptake of the muscle homogenate in the insect was determined during the paralysis which followed treatment with rotenone and methyl-parathion. It was recognized that the endogenous O_2-uptake of the homogenate was inhibited by addition of either rotenone or parathion. The dehydrogenase activity was depressed by parathion, but was increased by rotenone. The oxygen uptake of muscles and the degree of TTC staining of the nerve cord of the roach were depressed by in vitro treatment with 5×10^{-5} M rotenone, but were not influenced by 5×10^{-5} methyl parathion, 5×10^{-5} M DDT or 6.6×10^{-5} pyrethrine. From the accumulated data on the mode of action of rotenone it has been concluded that the primary action of rotenone in insects is the inhibition of the respiratory metabolism of nerve and muscle cells (Fukami 1956).

The effect of rotenone on respiratory enzymes of the nerve and muscle of *Periplaneta americana* was investigated further using mitochondrial fractions. All the six TCA cycle compounds tested were found to be oxidized. α-glycerophosphate, p-phenylendiamine and L-glutamate was also conspicuously oxidized, while pyruvate was, if any, only slightly oxidized. Rotenone at a concentration of 10^{-5}–10^{-7} M showed a remarkable inhibitory action on the oxygen consumption of the mitochondrial suspension in the presence of α-ketoglutarate, L-glutamate, malate, fumarate, oxalacetate, or citrate. It did not inhibit the oxidation of succinate, α-glycerophosphate, and p-phenylenediamine. From these results, it can be concluded that the action of rotenone is due to a specific inhibition of some oxidative enzymes with a pyridine nucleotide.

Effects of rotenone on the NADH oxidation were examined using a spectrophotometer. As is shown in these results, addition of cytochrome c raised the NADH oxidation activity of the cockroach muscle mitochondria, while antimycin A inhibitetd the oxidation. It is of interest to note that, while rotenone inhibited the NADH oxidation, dehydrorotenone, a closely related compound, did not inhibit the oxidation. Dehydrorotenone causes no in vivo toxicity (Fukami and Tomizawa 1958) (see Fig. 4).

Studies with rotenone and many derivatives have revealed a close correlation among the in vitro inhibition of L-glutamate oxidation, the blockage of nerve conduction, and in vivo toxicity in various insect species. In no case was there high insecticidal activity in the absence of enzyme inhibition (Fukami et al. 1959) (see Table 2).

The in vitro effect of rotenone on the oxygen uptake of the homogenate of nerve mitochondria with or without succinate were investigated using guinea pig's brain. Rotenone inhibited the endogenous oxygen, but did not inhibit the oxygen uptake, in the presence of succinate. Similar experiments using homogenates of the insect nervous system are also shown for comparison. It can be seen that these results demonstrate no substantial difference between insects and mammals (Fukami and Tomizawa 1958).

The same experiments have been conducted in the presence of L-glutamate. Rotenone, dihydrorotenone and rotenone-I inhibited the glutamate oxidation,

Table 2. Inhibition of L-glutamic dehydrogenase, block of nerve conduction and insecticidal action by rotenone derivatives. (Fukami et al. 1959)

No.	Compound	Insecticidal toxicity	Inhibition of glutamic dehydrogenase (%)	Time for block of nerve conduction (min) (No. of preparation)	Chemical structure			
					Chromano-chromanone ring	Asymmetric carbon atom (s) in: Dihydro-furan ring	Dihydro-pyron-pyran ring: Position 7	Dihydro-pyron-pyran ring: Position 8
1.	Rotenone	100	95 ~100	17(15)	+	+	+	+
2.	Dihydro-rotenone	75	92 ~100	18(6)	+	+	+	+
3.	Iso-rotenone	0	21.5~ 22.5	40(2), >80(4)	+	−	+	+
4.	Rotenone-hydrochloride	20	89 ~100	23(5), >60(1)	+	+	+	+
5.	Degueline	60	68 ~ 73	45(6)	+	−	+	+
6.	Tetrahydro-rotenone	0	0		+	−	+	+
7.	Rotenol (N)	50	75 ~ 84.5	26(6)	−	+	+	+
8.	Dihydro-rotenol (N)	20~35	68 ~ 85	47(5), >80(1)	−	+	+	+
9.	Acetyl-rotenone	50	70 ~ 74	25(6)	−	+	+	+
10.	Rotenone-oxim	0	15.5		−	+	+	+
11.	Rotenolone-I	20~30	45 ~ 56	16(6)	±	+	+	+
12.	Rotenolone-II	0	0 ~ 9.0	33(3), >60(6)	±	+	+	+
13.	Dehydro-rotenone	0	3.5	>60(6)	−	+	−	−
14.	Deguelinol-I	0	0		±	−	+	+
15.	Deguelinol-II	0	0		±	−	+	+
16.	Dehydro-rotenol (N)	0	8.0		−	+	−	−
17.	Rotenolol-II α (N)	0	0		−	+	+	+
18.	Rotenolol-II β (N)	0	0		−	+	+	+
19.	Rotenolol-Iα (N)	0	0		−	+	+	+
20.	Rotenolol-I β (N)	0	0		−	+	+	+
21.	Dihydrodehydro-rotenone	0	0		−	+	−	−
22.	Desoxy-dehydro-rotenone (N)	0	2.5		−	+	−	+
23.	Dihydro-desoxy-dehydro-rotenone (N)	0	13.0		−	+	−	+
24.	Derrisic acid	0	0	>60(6)	−	+	−	−
25.	Dihydro-derrisic acid	0	0	>60(6)	−	+	−	−
26.	Toxicaric acid	0	7.5~10.5	>60(6)	−	−	−	−

27.	(N)	0	0	−	+	−	−
28.	Tubaic acid	0	0	−	+	−	−
29.	(N)	0	0	−	−	−	−
30.	(N)	0	0	−	−	−	−
31.	(N)	0	0	−	−	−	−
32.	(N)	0	0	−	−	−	−
33.	(N)	0	4.5	−	−	−	−
34.	$C_{23}H_{22}O_6$ (N) (Isomer of rotenone)	0	0	−	−	−	−
35.	(N)	0	0	−	−	−	−

N: New compound

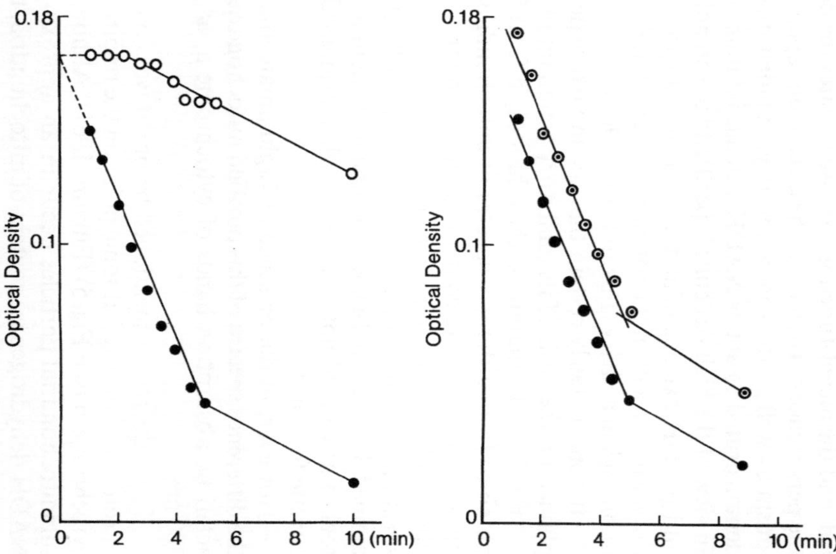

Fig. 4. The effects of rotenone and dehydrorotenone on the NADH-oxidation. The composition of the reaction mixtures were the same as shown in Fig. 3A. The final concentration of rotenone and dehydrorotenone: 0.42×10^{-5} M. –●–●– Control; –○–○– Rotenone; –⊙–⊙– Dehydrorotenone. (Fukami 1961)

Table 3. In vitro effects of rotenone and its derivatives on the O_2 uptake of the mitochondrial fractions of insect muscle and guinea pig brain in the presence of L-glutamate

Compound	Final concentration (M)	Inhibition of O_2 uptake (%)	
		Guinea pig brain	Insect muscle
Rotenone	10^{-6}	64	85
	10^{-6}	64	82
Dihydro-rotenone	10^{-5}	59	82
Rotenolone-I	10^{-5}	34	75
Rotenolone-II	10^{-5}	–	6
Dehydro-rotenone	10^{-5}	+18	14

but dehydrorotenone did not. As mentioned before, dehydrorotenone is not toxic. It can be seen that a similar tendency holds for both insects and mammals (see Table 3).

The metabolic pathway for rotenone is the same in each of the animals studied, and the mixed-function oxidase initiates the detoxication reactions and limits the persistence of rotenone. These reactions appear to be the detoxification mechanism as well as the source for as yet unidentified metabolites which have been demonstrated to act as NADH oxidase inhibitors and also as toxicants for mice. In view of the high structural specificity of rotenoids, with respect to both toxicity and NADH oxidase inhibition, it is not surprising that the structural modifications which occur in the rotenone molecule during metabolism cause a reduction in biological activity (Fukami et al. 1969).

It appears fairly certain that the in vivo toxicity of rotenone can be fully explained on the basis of its inhibitory action on the respiratory chain.

Further information on the inhibitory effect of rotenone and antimycin A was obtained from difference spectra of mitochondrial preparations using cockroach muscle mitochondria containing the inhibitor and L-glutamate, α-glycerophosphate, and NADH under various conditions. The effect of antimycin A on the absorption spectrum of reduced cytochromes of mitochondrial suspensions was examined by the opal-glass transmission method. The absorption band of cytochrome c could not be detected when examined immediately after shaking the cuvette for some time.

However, in the presence of L-glutamate and after treatment with rotenone the difference spectra of this purified mitochondrial preparation showed the characteristic absorption bands of cytochrome a, a^3, b and c between 330 nm and 360 nm.

From the results mentioned before, it was concluded that the inhibitory effect of rotenone is exerted at some point on the respiratory chain between NADH and cytochrome b (see Fig. 5) (Fukami 1961). Although specific rotenone binding to submitochondrial particles occur in an approximately 1:1 molar ratio with the NADH dehydrogenase content of mitochondria, the extraction of phosphorylat-

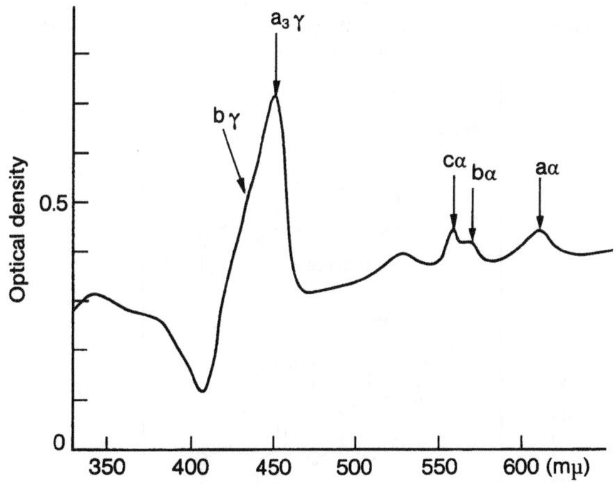

Fig. 5. Difference spectrum of rotenone in purified mitochondrial suspension containing L-glutamate (Fukami 1961). In the test cuvette was 2.7 ml of reaction mixture containing 1 ml of mitochondrial suspension (1.52 mg N ml^{-1}), 0.3 ml of 0.5 M L-glutamate and rotenone (5×10^{-5} M). The control cuvette was identical except for the addition of rotenone. Temperature 19 °C

ing or nonphosphorylating particles labled with [^{14}C]-rotenone results in almost complete separation of [^{14}C]-rotenone from this enzyme. In addition, rotenone displaces a portion of the endogeneous ubiquinone involved in electron transport from NADH but not from succinate. It, therefore, appears that the inhibition of NADH-linked respiration by rotenone occurs through specific binding to a component of the respiratory chain between NADH dehydrogenase and coenzyme Q and not to NADH dehydrogenase itself (Fukami 1976, 1984).

During the course of a screening program to discover new types of insecticides from microorganisms two materials, piericidin A and B were isolated from *Streptomyces mobaraensis* and their chemical structure has been elucidated (see Fig. 6). These compounds were found to have a broad spectrum of insecticidal activity. The inhibitory activity of piericidin A toward the aerobic oxidation of NADH-linked substrates was similar to that of rotenone. The strong structural resemblance of piericidin A to coenzyme Q and the fact that the inhibition of succinate oxidation of piericidin A could be reversed by the addition of coenzyme Q strongly suggested that piericidin A was acting as a competitive inhibitor of coenzyme Q (Mitsui et al. 1969).

Yet another compound apart from rotenone and piericidine has the same mode of action: papaverine (Fukami 1976, Tsuda et al. 1977) (see Fig. 6). Its chemical structure is similar to that of rotenone. But papaverin does not show its insecticidal potency in vivo because its structure contains polar substituents and, therefore, high water solubility (Fukami 1976).

We are very much interested in the bypass of the respiratory chain by rotenone and piericidin. Conover and Ernster (1960) concluded that vitamin K_3 induces a bypass of the rotenone-sensitive site of the respiratory chain through DT-diaphorase. As shown in these results the vitamin K_3 effect can also be found in liver mitochondria after the addition of rotenone and piericidin A. In sharp difference to rat liver mitochondria, cockroach muscle mitochondria show no restoration of respiration after the addition of vitamine K_3 to the system inhibited by piericidin A or rotenone. The other quinones, such as vitamin K_1, benzoquinone, Co Q_9 and Co Q_{10} have no effect in rat liver mitochondria. Similar phenomena

Structure	Inhibition (%)	Structure	Inhibition (%)
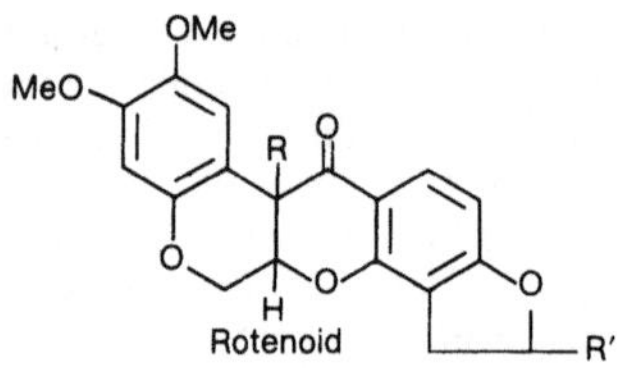 Rotenoid Rotenone oxidative metabolites Rotenone: R = H, R′ = $-C(=CH_2)CH_3$ Rotenolone I R = OH Rotenolone II R = OH 8′-Hydrooxyrotenone: R = $-C(=CH_2)CH_2OH$ 6′-7′-Dihydro-6′, 7′-dihydroxyrotenone : R = $-C(CH_2OH)(OH)CH_3$	 100 80 0 80 100	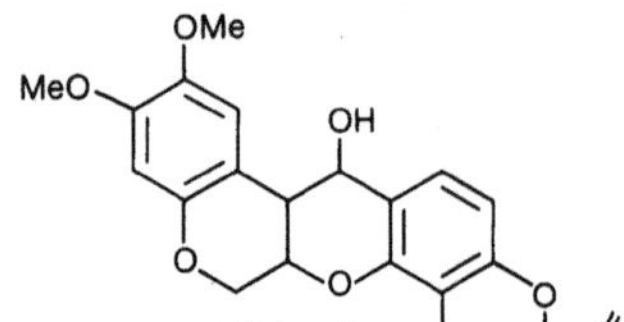 Rotenol 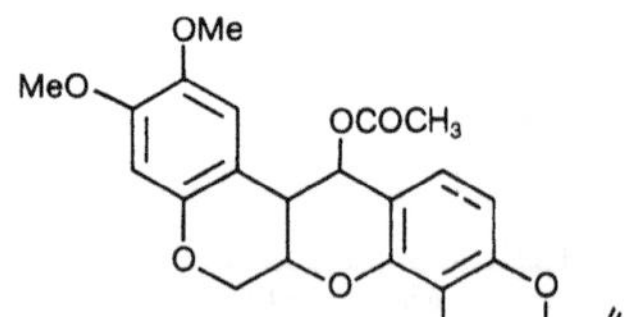 Acetylerotenone 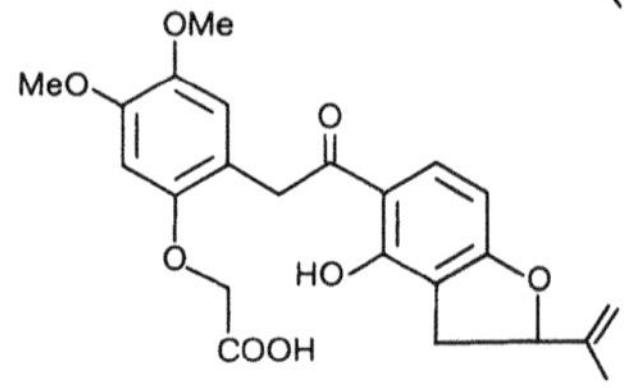	50 40 0
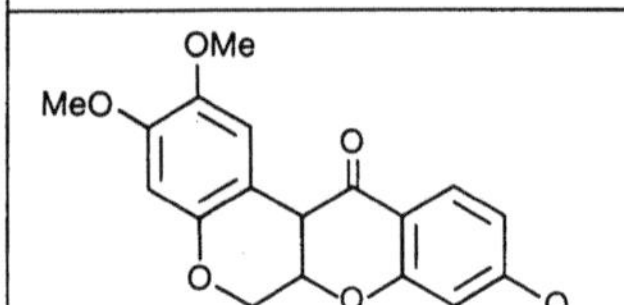 Degueline	60	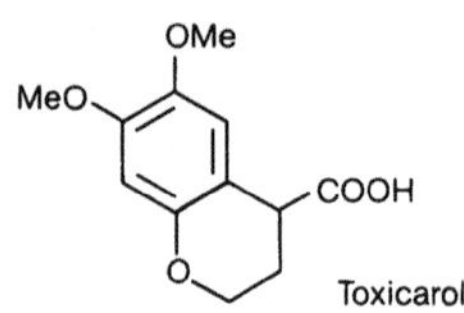 Toxicarol	0
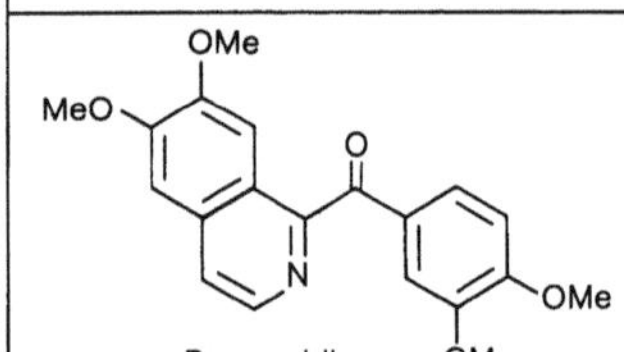 Papaveridine	40	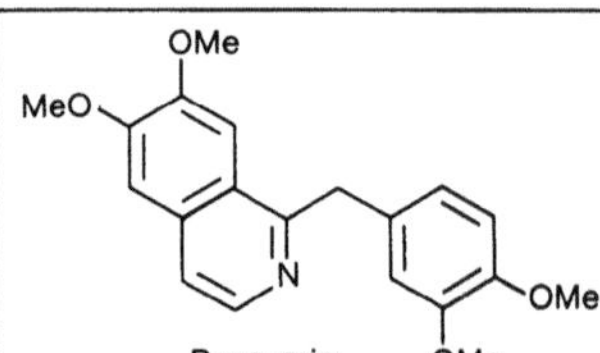 Papaverin	50
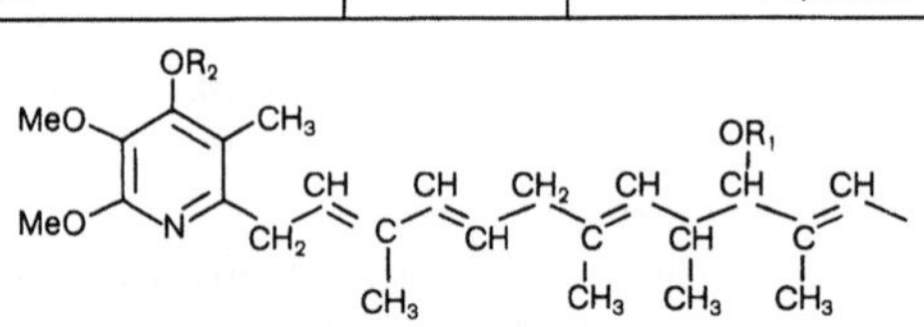 Piericidin A: $R_1 = H_3$ $R_2 = H$			100
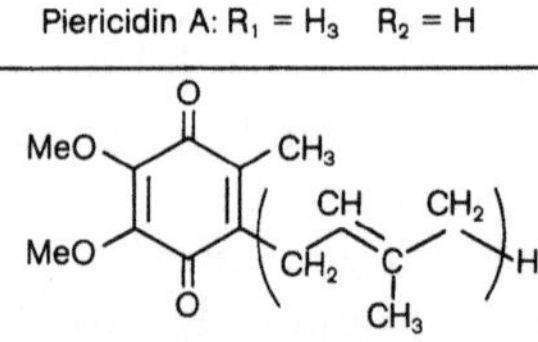 Co-enzyme Q			

Fig. 6. The effects of rotenoid, piericidin and papaverin on NADH oxydase of rat liver mitochondria. (Fukami 1976, Tsuda et al. 1977)

are found in other mitochondria from cockroach mid-gut and honey bee muscle. Specific and total activities of NADH and NADPH-diaphorase in rat liver are highest in the soluble fraction. Furthermore, NADH-diaphorase activity is generally higher than NADPH diaphorase in each fraction except for the soluble fraction. On the other hand, the activity of NADH- and NADPH-DT diaphorases from the American cockroach are found to be hardly different in each enzyme source, but generally they are lower than those of rat liver, which are $^1/_7$ and $^1/_3$ to $^1/_0$, respectively, in specific activity in each mitochondrial fraction (Mitsui et al 1970).

The conclusion from these results, therefore, is that it is very important to study (a) the physiological significance of the activity of vitamine K_3 on the bypass of the respiratory chain when it is inhibitetd by rotenone and (b) the DT diaphorase activity in order to understand the selectivity of the toxicity of rotenone between mammals, fish and insects.

3.1.2 Inhibitors of Cytochrome c-Oxidase

Apart from the inhibitors acting between NAD and coenzyme Q or those acting on the α-glycerophosphate shuttle, several inhibitors of succinate and inhibitors acting between cytochrom b and c, have been found. However, attractive examples were investigated among the inhibitors of cytochrome c oxidase. Like CO, NO, N_3 and CN, the gas phosphine (hydrogen phosphide, PH_3) is a respiratory poison which is used as fumigant for the control of insect pests in stored products. Some studies have been conducted to explain the toxic effects of phosphine by its inhibition of the cytochrome c oxidase. Cytochrome components of mitochondria in the maize weevel, *Sitophilus zea mais* (M) were determined by the method of reduced minus oxidized difference spectra. When PH_3 (about 10 mM) was added to a suspension of the aerobic maize weevel mitochondria in a small cuvette at room temperature, clear peaks of α-cytochrome a, α-cytochrome c and γ-cytochrome a were visible with their characteristic time course. No spectral change was observed in the region of the peak of cytochrome b. This observation clearly suggested that PH_3 acts on the cytochrome oxidase. However, each peak position of the α-band (600 nm) and the γ-band (445 nm) of cytochrome a with PH_3 was significantly shifted to a different wave length compared with those obtained after HCN-treatment (Nakakita 1976) (see Fig. 7).

While this effect appears to be only slowly reversible in vitro, there is evidence that little inhibition of this enzyme occurs in vivo. The results show the effect of phosphine and cyanide on cytochrome oxidase when applied in vitro to mitochondrial preparations of *Rhyzopertha dominica.* It is clear that cyanide, a well-established cytochrome c-oxidase inhibitor totally abolished the activity in these preparations at 0.1 mM concentrations. Similarly phosphine, a postulated cytochrome oxidase inhibitor, produced 90–92% inhibition of the activity (see Fig. 8). When insects were treated with gaseous phosphine prior to isolation of mitochondria, only a small inhibitory effect on the cytochrome oxidase could be detected in the reference strain even though knock down has occurred and the insects were inactive prior to the isolation of their mitochondria. No inhibition of

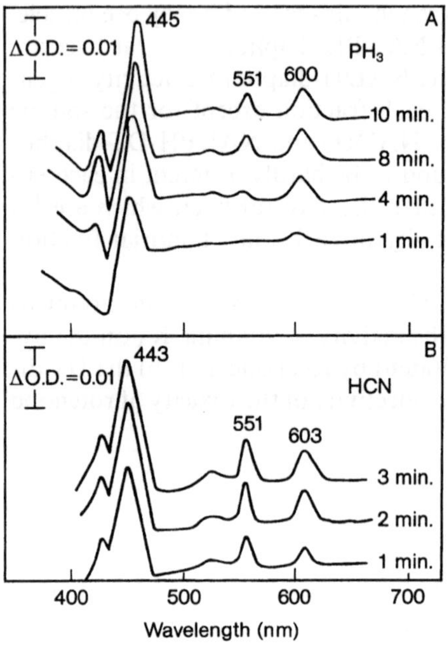

Fig. 7 A, B. Absorption spectra of PH_3 (**A**) and *HCN* (**B**) on the maize weevil mitochondria at room temperature. The inhibitor was added to the mitochondrial suspension of maize weevil in the sample cuvettes. Its absorption spectra were recorded against the suspension containing only mitochondria in a reference cuvette which was aerated before measurements. The concentration of mitochondrial protein was 2.8 mg ml^{-1}. (Nakakita 1976)

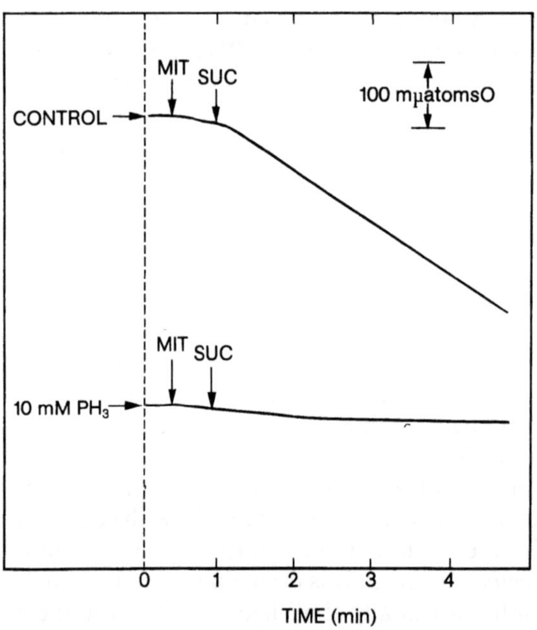

Fig. 8. Effect of PH_3 on respiration of sonicated mitochondria. In this experiment, sonicated mitochondria were used in place of intact mitochondria. Sonicated mitochondria (2 mg protein ml^{-1}), succinate (8 mM) and NAD (20 μM) were added at the points indicated by *arrows*. (Nakakita 1976)

activity was detected in the selected strain 306 when the physical activity of the insects prior to isolation of mitochondria was normal (Price 1980). The possibility that phosphine had been removed from the site of action during the isolation of the mitochondria was considered and the effect of phosphine on the cytochrome oxidase was compared in individual *R. dominica* before and after treatment.

Table 4. Effect of the exposure of *R. dominica* to 1 mg l^{-1} phosphine for 5 h on catalase activity. (Price et al. 1982)

	Catalase activity (μmol H_2O_2 reduced mg^{-1} protein min^{-1})	Inhibition (%)
Reference control	7.17 ± 0.2	60
Reference treated	2.9 ± 0.02	
306 control	13.69 ± 0.07	20
306 treated	10.89 ± 0.08	
KMI control	9.16 ± 0.08	12
KMI treated	8.04 ± 0.19	

These results show the effect of fumigating insects with HCN and PH_3 on their cytochrome c oxidases which were obtained from rapid homogenates of individual *R. dominica*. HCN treatment, resulting in knock down of both strains of *R. dominica* also produced almost total abolition of cytochrome oxidase activity, in the homogenate. With phosphine treatment no inhibition of the cytochrome c-oxidase activity was detected in homogenates of the selected strain 306 although phosphine produced knock-down in the reference strain (Price 1980).

Price et al. (1982) reported that phosphine inhibits catalase activity in *R. dominica*. As is shown in Table 4 catalase activity in reference insects is severly depressed after the insects had been exposed to 1 mg l^{-1} phosphine for 5 h. Inhibition of the enzyme in strain 306 is considerably reduced and in the KMI strain there is very little inhibition of catalase by phosphine. In addition, both 306 and KMI insects exhibit a higher control titre of the enzyme than do the reference insects. In the case of strain 306 this increase is about 90% and in the case of the strain KMI, about 30%. Uptake of phosphine and inhibition of catalase are also lower in resistant than in a susceptible strain (see Table 4).

Barsa and Ludwig (1959) reported that DDT at 10^{-3} M has a specific effect on cyanide sensitive respiration, in that it inhibits only cytochrome c-oxidase and succinate oxidation but has no effect on alcohol-, glucose-, lactic-, α-glycerophosphate- or isocitric-dehydrogenase in homogenates of the yellow mealworm and the housefly. However, the fact that DDE and other nontoxic DDT analogues also inhibit succinate oxidation in housefly homogenates led Anderson et al. (1954) to conclude that this inhibition was not a prime factor in the mode of action of DDT.

3.1.3 The Inhibitors of the α-glycerophosphate Shuttle

The α-glycerophosphate dehydrogenase activity is very high in the α-glycerophosphate shuttle, especially, as mentioned earlier, in the flight muscles of insects. Marquardt and Brosemer (1966) compared the pattern of isozyme on the glycerophosphate dehydrogenase between flight muscles of the bee and the skeletal

Table 5. The effects of cinnamic acid and its substituted compounds (HO— —CH=CHCOOH) on the α-glycerophosphate dehydrogenase from the housefly flight muscle (O'Brien et al. 1985)

Substitutional group	Concentration (M)	Inhibition (%)
p-Nitro	10^{-4}	60
	5×10^{-5}	21
	2×10^{-5}	0
p-Hydroxy	10^{-4}	40
m-Hydroxy	10^{-4}	40
o-Hydroxy	10^{-4}	26
m-Chloro	10^{-4}	72
p-Methyl	10^{-4}	42

muscles of the rabbit. The pattern of isozyme is apparently different between them. O'Brien et al. (1965) investigated the effect of ciannamic acid and its analogues on the α-glycerophosphate dehydrogenase in the flight muscle of the housefly. As can be seen in Table 5, compounds with a p-nitro substitution inhibited only 21% at 5×10^{-5} M, which is considered to be a physiological concentration.

We investigated the effects of several compounds such as ciannamic analogues on the α-glycerophosphate oxidation in the housefly. Among 150 compounds only 2,4-dinitrophenolthiocyanate showed 55% inhibition of the α-glycerophosphate dehydrogenase at 5×10^{-5} M, but it had no insecticidal activity. For comparison, rotenone showed 50% inhibition of NADPH oxidase in housefly mitochondria at a concentration of 5×10^{-7} M.

3.2 Uncoupling Agents

Under normal conditions the mitochondrial respiratory chain does not function independently but is closely linked to and controlled by the reactions associated with oxidative phosphorylation. Thus, as has been described before the addition of a respiratory substrate to an intact ("coupled") mitochondrial preparation increases the rate of respiration (state 3) until the supply of ADP and Pi is exhausted through conversion to ATP. In the presence of an agent such as 2,4-dinitrophenol, the close coupling between the respiratory chain and phosphorylation is broken, respiratory control is lost, and electron transport along the chain occurs at a maximal rate without formation of ATP. Compounds such as 2,4-dinitrophenol are therefore called "uncoupler".

Dinitrophenols such as the miticides DNOC (4,6-dinitro-0-cresol) and DNCHP (2,4-dinitro-6-cyclohexylphenol) are well-established uncouplers of oxidative phosphorylation and their activity is enhanced by a low pKa and high degree of lipid solubility. Salicylanilides have recently been shown to possess important fungicidal, bacteriostatic, cestocidal, and molluscicidal properties, and 5-chloro-3-t-butyl-2-chloro-4-nitrosaylicylanilide has been developed for the control of schistosomiasis (Whitehouse 1964). These compounds showed uncoupling

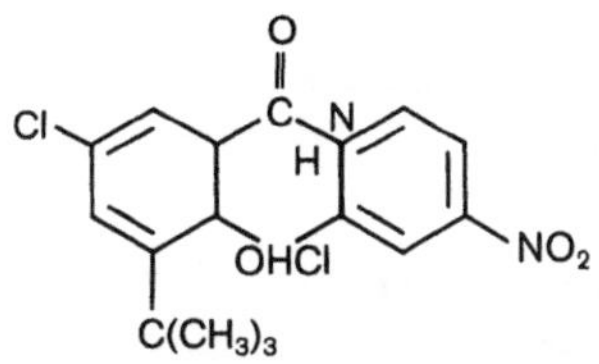

5-Cl-3-*t*-butyl-2′-Cl-4′-NO_2-salicylanilide

Fig. 9. The chemical structure of 5-Cl-3-t-butyl-2-Cl-4-NO_2-salicylanilide

activity on the oxidative phosphorylation at 10^{-9} M (Fig. 9). The concentration of the compounds is very low, because agricultural chemicals such as dinitrophenol show their uncoupling activity only at concentrations from 10^{-5} to 10^{-6} M. In general, mitochondria from rat liver appear to be less sensitive to the action of uncoupling agents than those from mouse brain or from insect tissues (Table 6).

Table 6. The effects of phenols and other chemicals on the respiratory system compared with their toxicity to the American cockroach[b]

Chemical	Conc. (mM)	American cockroach		Rat liver		Insecticidal toxicity
		Stimulation of state 4 respiration	% of P/O	Stimulation of state 4 respiration	% of P/O	
Untreated	–	0	100.0	0	100.0	0
DNP	0.1	40	84.6	50	84.0	60
	1.0	130	23.1	180	0.0	
o-Nitrophenol	1.0	– 30	83.1	0	97.0	0
	10.0	– 10	47.1	20	94.1	
p-Nitrophenol	1.0	20	23.1	90	68.5	0
	10.0	1,030	0.0	30	0.0	
3-Methyl-4-nitrophenol	1.0	50	61.5	100	56.6	20
	10.0	790	38.5	110	0.0	
o-Chlorophenol	1.0	10	83.1	10	97.0	0
	10.0	70	93.5	0	94.1	
m-Chlorophenol	1.0	10	64.5	0	89.3	20
	10.0	260	65.7	100	19.1	
p-Chlorophenol	1.0	10	96.4	0	48.4	0
	10.0	270	78.2	90	10.7	
2,4-Dichlorophenol	1.0	70	89.3	60	54.2	60
	10.0	490	0.0	20	0.0	
PCB	0.01	10	95.8	40	77.8	
	0.1	70	50.0	160	11.1	
p-Cyanophenol	1.0	300	80.2	20	44.6	0
	10.0	– 10	77.8	130	16.7	

[a] 1 γ

[b] Pyruvate and succinate were used as substrates for oxidations by cockroach and rat mitochondria respectively. The chemicals (50 γ) were administered by intrathoracic injection to the cockroach

Table 6. (continued)

Chemical	Conc. (mM)	American cockroach		Rat liver		Insecticidal toxicity
		Stimulation of state 4 respiration	% of P/O	Stimulation of state 4 respiration	% of P/O	
Benzylalcohol	1.0	– 20	93.6	– 10	58.9	20
	10.0	– 10	100.0	– 10	60.7	
Benzylmercaptan	1.0	10	83.5	90	69.1	20
	10.0	50	85.5	– 50	28.6	
o-Cresol	1.0	10	98.4	10	83.1	0
	10.0	40	100.0	0	80.0	
m-Cresol	1.0	10	100.0	0	86.8	0
	10.0	50	100.0	10	88.0	
p-Cresol	1.0	10	56.2	– 10	91.6	20
	10.0	30	98.0	0	86.6	
3-Nitro-4-cresol	1.0	0	99.5	0	81.9	40
	10.0	70	61.7	90	60.2	
2,6-Di-*tert*-butyl-cresol	1.0	– 10	92.5	0	88.0	40
	10.0	– 50	0.0	– 60	8.4	
3,4-Xylenol	1.0	– 20	82.1	0	83.1	0
	10.0	30	88.1	20	71.1	
o-Aminophenol	1.0	– 30	93.6	– 10	82.7	0
	10.0	– 20	94.0	10	63.1	
m-Aminophenol	1.0	– 20	99.5	– 10	77.1	0
	10.0	0	100.0	– 30	86.8	
p-Aminophenol	1.0	– 20	91.9	0	79.5	20
	10.0	– 10	88.7	60	92.3	
4-(Methylthio) phenol	1.0	– 10	91.5	0	84.5	100
	10.0	50	87.5	60	69.7	
Methyl parathion	1.0	0	100.0	0	95.8	100[a]
	10.0	– 80	–	0	56.6	
Sumithion	1.0	– 20	98.0	0	89.9	100[a]
	10.0	– 70	–	0	45.8	
Lindane	1.0	– 10	37.9	– 30	84.5	100[a]
	10.0	– 100	–	– 50	0.0	

3.3 Inhibitors of the Energy Transfer System

Reports on inhibitors of the energy transfer system have appeared only relatively recently. Hollunger (1955) reported that guanidine inhibits only respiration coupled with phosphorylation (state 3) and that it has no effect on the respiration of uncoupled mitochondrial particles. Subsequently, oligomycin was found to inhibit energy transfer and it is the best known compound of this type. Oligomycin exhibits properties which are different from those of uncoupling agents. It causes little or no inhibition of state 4 respiration. Respiration in the presence of oligomycin is not stimulated by the addition of ADP but it increases remarkably after the addition of 2,4-dinitrophenol. The addition of oligomycin to mitochondria respiring in state 3 (after addition of ADP) causes an immediate inhibition of respiration to the level of state 4. The characteristics of compounds inhibiting energy transfer are that (a) they inhibit state 3 respiration and ATP synthesis, (b) they inhibit latent ATPase activity stimulated by uncoupling agents, (c) they inhibit the ATP-Pi exchange reaction, and (d) they inhibit the ATP-ADP exchange reaction.

Of several types of compounds with a mode of action similar to that of oligomycin, the organotins are the only ones of any insecticidal importance. Organotins are becoming increasingly important as miticidal and biocidal agents, and the best known representatives of this group, are the trialkyltins. These materials act directly on the energy conservation processes in the mitochondria of both insects and mammals.

3.4 Others

It has long been acknowledged that the toxicity of DDT on the nervous system is due or related to disruptions of Na^+ and K^+ transport processes in the nerve membrane. Recently it has been shown that various chlorinated hydrocarbon insecticides including DDT may inhibit nerve ATPase in vitro. To determine the I_{50} values for DDT and DDE inhibition using the Na^+-K^+-Mg^{++}-ATPase from the nerves of walking leg of the lobster, the enzyme in the standard assay buffer was incubated with various concentrations of DDT and DDE. These results show the concentration-inhibition relationship for DDT and DDE. The I_{50} values for DDT (3×10^{-6} I and DDE (3×10^{-5}) can be seen in Table 7. Both, Na-K-Mg ATPase and the electrical activity in a nerve poisoned with other pharmacological agents are shown. To study the relationship between the symptoms of poisoning in nerve cells and the inhibition of ATPases further, several chemicals of pharmacological interest were tested, and compared with some data collected from the literature. DDT has been demonstrated to affect the ionic conductance of the lobster nerve membrane. The major effects are the slowing of the sodium inactivation and the suppression of the potassium conductance increase. These changes in conductance can account for the prolongation of the action potential and the latter initiates repetitive after-discharge. Tetrodotoxin (TTX) was without effect on the ATPase activity. The toxin is one of the most effective nerve blocking agents and inhibits the mechanism of sodium conductance increase without affecting other conductance mechanisms. However, it is totally ineffective when applied to the inner surface of a cell membrane (Matsumura and Narahashi 1971). Various ATPases of the synaptic region of the rat brain were tested for their susceptibility to heptachlor repoxide. As is shown in Table 8, it became clear from these results that Na^+, K^+-ATPase, no matter what nerve components were studied, did not show appreciable sensitivity to these insecticides. On the other hand, CA^{2+}, Mg^{2+}-ATPase showed the highest level of inhibition among all ATPases tested. The high level of inhibition by heptachlor epoxide on the enzyme activity was observed in crude synaptosomal membranes. It is known that this fraction contains synaptic membrane components. These membrane preparations also showed a

Table 7. The effects of various pharmacological agents on the electrical ATPase activities of nerve fibers. (Matsumura and Narahashi 1971)

Chemicals	ATPase activity I_{50} *in vitro*	Electrical activity			
		Minimum effective concn[a]		Action potential[c]	Ionic conductances inhibited[d]
		Externally	Internally[b]		
DDT	3×10^{-6} M	3×10^{-6} M		NAP, RD	$g_{Na}\downarrow$, $g_K\uparrow$
DDE	3×10^{-5} M	1×10^{-5} M		No effect	
Hemicholinium-3	5×10^{-5} M	1×10^{-2} M[e]	1×10^{-3} M[e]	Block	$g_{Na}\uparrow$, $g_K\uparrow$
Tetraethylammonium	1×10^{-3} M	3×10^{-1} M[e]	6×10^{-4} M[e]	NAP	$g_K\uparrow$[17,18]
N-acetyl imidazole	5×10^{-5} M	5×10^{-4} M[f]		NAP, RD[g]	
Veratrine	1×10^{-4} g ml^{-1}	2.4×10^{-5} g ml^{-1}[e]		NAP, RD	
Tetrodotoxin	$>2 \times 10^{-4}$ M	1×10^{-8} M[e]	$>1 \times 10^{-5}$ M[e]	Block	$g_{Na}\uparrow$
Grayanotoxin I	$>1.2 \times 10^{-3}$ M	2.4×10^{-4} M[f]		NAP, RD	None

[a] The lowest concentration range that causes the definite pharmacological effects on the electrical activity of nerve
[b] Data with internally perfused squid giant axons
[c] NAP, negative after-potential is increased and prolonged; RD, repetitive discharges are produced
[d] $g_{Na}\uparrow$, the mechanism by which the membrane sodium conductance is increased upon depolarization; $g_{Na}\downarrow$, the mechanism by which the membrane sodium conductance is decreased during depolarization; $g_K\uparrow$, the mechanism by which the membrane potassium conductance is increased upon depolarization
[e] Data with squid giant axons
[f] Data with crayfish giant axons
[g] W.R. Kem, personal communication

Table 8. The effects of heptachlor epoxide and γ-BHC on ATPase activity. (Yamaguchi et al. 1979)

ATPase	Fraction used	Specific activity (μmol *P* mg^{-1} 30 min^{-1})	Percentage inhibition[a]	
			Hepta-chlor epoxide	γ-BHC
Na^+, K^+-ATPase	Synaptosomal	4.9 ± 0.2	0	0
(Ouabain-sensitive)	Microsomal	3.6 ± 0.2	0	0
Na^+, K^+-ATPase[b]	Synaptosomal	2.3 ± 0.3	7.8 ± 2.1	4.1 ± 1.1
(Ouabain-insensitive)	Microsomal	3.7 ± 0.2	18.4 ± 2.8	2.8 ± 1.5
Ca^{2+}-ATPase	Synaptosomal	1.3 ± 0.2	47.8 ± 2.0	16.4 ± 0.8
	Microsomal	2.5 ± 0.4	48.0 ± 7.2	25.0 ± 8.4
	Synaptic vesicle	3.4 ± 0.2	24.4 ± 1.8	1.5 ± 0.8
	Synaptic membrane	1.3 ± 0.3	26.5 ± 2.0	0
Mg^{2+}-ATPase	Synaptosomal	1.2 ± 0.3	31.2 ± 5.1	11.4 ± 1.3
	Microsomal	3.7 ± 0.4	53.4 ± 4.9	4.1 ± 1.0
	Synaptic vesicle	6.2 ± 0.9	47.1 ± 9.6	15.5 ± 1.6
	Synaptic membrane	4.4 ± 1.1	48.9 ± 8.5	0
Ca^{2+}, Mg^{2+}-ATPase	Synaptosomal	0.4 ± 0.1	66.8 ± 11.8	32.4 ± 5.3
	Microsomal	1.3 ± 0.4	72.7 ± 5.2	45.3 ± 8.7
	Synaptic vesicle	1.8 ± 0.2	79.5 ± 8.8	42.9 ± 4.1
	Synaptic membrane	1.1 ± 0.2	76.9 ± 4.2	19.9 ± 11.2

[a] Data are expressed as percentage inhibition of control (without inhibitor), where the inhibitor concentration is 10^{-5} *M* and the protein content of the preparations is 20–50μg tube^{-1}. Each value is the mean ± standard deviation from three to four determinations
[b] The residual Na^+, K^+-ATPase activity in the presence of ouabain

Table 9. The effects of in vivo administered heptachlor epoxide on synaptosomal (Ca^{2+}-Mg^{2+}-ATPase activity and its distribution in synaptosomal subfractions[a]. (Yamaguchi et al. 1980)

Fractions	Ca^{2+}-Mg^{2+}-ATPase activity[b] (nmoles $^{32}P_i\,mg^{-1}$ protein $20\,min^{-1}$) Control	Treated	Distribution of heptachlor epoxide (ng/mg protein)
1	150± 42	0[c]	160
2	504± 98	64± 48[c]	199
3	1,775±134	1,642±186	156
4	1,563± 96	2,054±100	130
5	286±102	292± 78	33

[a] Heptachlor epoxide dissolved in corn oil was intraperitoneally injected into the rats ($200\,mg\,kg^{-1}$). When convulsions were observed, the rats were killed and the synaptosomes were prepared by the method of Hajos. Thy synaptosomes were disrupted and fractionated as in the legend of Table 6. Ca^{2+}-Mg^{2+}-ATPase activity was assayed in a medium containing 100 mM KCl, 3 mM $MgCl_2$, 0.1 mM ouabain, 30 mM imidazole-HCl, pH 7.2, and EGTA-equilibrated Ca^{2+}, 10^{-8} M (basal medium) or 5×10^{-6} M (medium for activation). [^{32}P]ATP (0.5 mM) was used as substrate. The determination of the level of heptachlor epoxide in the brain and subcellular fractions was carried out by gas liquid chromatography after extraction by ether

[b] Expressed as the average of two independent experiments (each experiment involving two duplicate determinations) ± standard error

[c] Difference against control significant at <0.005

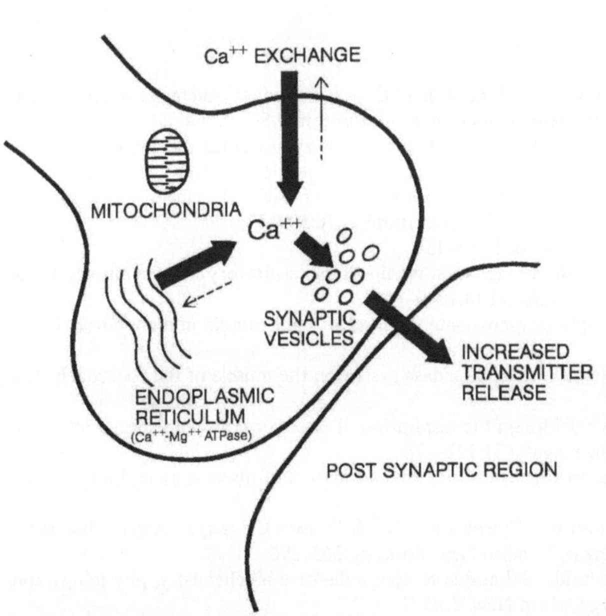

Fig. 10. Schematic illustration of the overall effects of heptachlor epoxide on the processes of transmitter release. *Arrows* represented by *solid lines* show that the process is stimulated by heptachlor epoxide. Those represented by *dotted lines* show its inhibitory effects. The overall effect of heptachlor epoxide is the increase of internal free Ca^{2+} which triggers the release of the transmitter. The role of mitochondria in Ca^{2+} regulation was not examined in this work. (Yamaguchi et al. 1980)

high titer of CA^{2+} -Mg^{2+} ATPase sensitive to these insecticides (Yamaguchi et al. 1979) (Table 8). The effect of in vivo administered heptachlor epoxide on synaptosomal Ca^{2+} -Mg^{2+} ATPase activity and its distribution in synaptosomal subfractions using rat brain were investigated. As is shown in these results, it was found that the Ca^{2+} -Mg^{2+} ATPase of the same two fractions which had previously been shown to be susceptible to heptachlor epoxide poisoning in vitro inhibited also in vivo (Table 9). In summary, it has been established (Yamaguchi et al. 1980) that heptachlor epoxide induces an elevation of intracellular CA^{2+} in the synaptosomes. This increase most likely plays a significant role in the heptachlor epoxide induced increase of transmitter release. The current working hypothesis attempting to explain this intracellular Ca^{2+} increase states that heptachlor epoxide inhibits CA^{2+} -Mg^{2+} ATPase. This has been corroborated by in vitro experiments (Fig. 10).

4 Conclusion

Assessing the results from these experiments with inhibitors of respiration we realize that they have been obtained to a large extent from muscle and fat body tissues. Yet the primary toxic effects of insecticides are considered to be exerted on nerve cells. Therefore, the physiology of nerve cell respiration is very important. In order to understand the selectivity of insecticides between vertebrates and invertebrates we have to study the activity of these compounds using nerve cell preparations, too. From such studies, I trust, we will be able to develop new types of insecticides with low mammalian toxicity.

References

Anderson AD, March RB, Metcalf RL (1954) Inhibition of the succinoxidase system of susceptible and resistant houseflies by DDT and related compounds. Ann Entomol Soc Am 47:567–595

Barsa MC, Ludwig D (1959) Effects of DDT on the respiratory enzymes of the mealworm, *Tenebrio molitor* (L.), and of the housefly, *Musca domestica* (L.). Ann Entomol Soc Am 52:179–185

Chance B, Sacktor B (1958) Respiratory metabolism of insect flight muscle. II. Kinetics of respiratory enzyme in flight muscle sarcosomes. Arch Biochem Biophys 76:509–531

Chance B, Williams GR (1956) Adv Enzymol 17:65–134

Conover T, Ernster L (1960) Bypass of the amytal-sensitive site of the respiratory chain in mitochondria by means of vitamin K_3. Acta Chem Scand 14:1840–1848

Estabrock RW, Sacktor B (1958) α-glycerophosphate oxidase of flight muscle mitochondria. J Biol Chem 233:1014–1019

Fukami J (1954) Effect of rotenone on the succinoxidase system in the muscle of the cockroach. Jpn J Appl Zool 19:29–37

Fukami J (1956) Effect of some insecticides on the respiration of insect organs, with special reference to the effects of rotenone. Botyu-Kagaku 21:122–128

Fukami J (1961) Effect of rotenone on the respiratory enzyme system of insect muscle. Bull Natl Sci (Wellington) C 13:33–45

Fukami J (1976) Insecticides as inhibitors of respiration. In: Wilkinson CF (ed) Insecticide biochemistry and physiology. Plenum Press, London New York, pp 363–396

Fukami J (1984) Rotenone and rotenoids. Submitted to comprehensive biochemistry, physiology and pharmacology. Pergamon Press, Oxford New York

Fukami J, Tomizawa C (1958) Effect of rotenone and its derivatives on the respiration of brain in guinea pig. Botyu-Kagaku 23:205–208
Fukami J, Nakatsugawa T, Narahashi T (1959) The relation between chemical structure and toxicity in rotenone derivatives. Jpn J Appl Entomol Zool 3:259–265
Fukami J, Shishido T, Fukunaga K, Casida JE (1969) Oxidative metabolism of rotenone in mammals, fish and insects and its relation to selective toxicity. J Agric Food Chem 17:1217–1226
Hollunger G (1955) Guanidines and oxidative phosphorylation. Acta Pharmacol Toxicol Suppl No 1 11:84
Ilivicky J, Chefurka W, Casida JE (1967) Oxidative phosphorylation and sensitivity to uncouplers of housefly mitochondria: Influence of isolation medium. J Econ Entomol 60:1404–1409
Jeng M, Hals C, Crane FL, Takahashi S, Tamura S, Folkers K (1968) Inhibition of mitochondrial electron transport by piericidin a and related compounds. Biochemistry 7:1311–1317
Lardy H, Ferguson SM (1969) Oxidative phosphorylation: Role of inorganic phosphate and acceptor system in control of metabolic rates. J Biol Chem 195:215–222
Marquardt RR, Brosemer RW (1966) Insect extramitochondrial α-glycerophosphate dehydrogenase. I. Crystallization and physical properties of the enzyme from honeybee (Apis ellifera) thoraces. Biochim Biophys Acta 128:454–460
Matsuda M, Fukami J (1972) Preliminary survey of effects of phenols on the oxidative phosphorylation in the american cockroach muscle mitochondria. Appl Entomol Zool 7:27–36
Matsumura F, Narahashi T (1971) ATPase inhibition and electrophysiological change caused by DDT and related neuroactive agents in lobster nerve. Biochem Pharmacol 20:825–837
Mitsui T, Fukami J, Fukunaga K, Takahashi N, Tamura S (1969) Studies on piericidin. I. Effect of piericidin a and b on the mitochondrial electron transport in insects. Butyu-Kagaku 34:135–139
Mitsui T, Fukami J, Fukunaga K, Takahashi N, Tamura S (1970) Studies on piericidin: Antagonistic effect of vitamin K_3 on the respiratory chain of insects and mammals in the presence of piericidin. J Agric Biol Chem 34:1101–1109
Nakakita H (1976) The inhibitory site of phosphine. J Pestic Sci 1:235–238
O'Brien RD, Cheng L, Kimmel EC (1965) Inhibition of the α-glycerophosphate shuttle in housefly flight muscle. J Insect Physiol 11:1241–1248
Price NR (1980) Some aspects of the inhibition of cytochrome c oxidase by phosphine in susceptible and resistant strains of *Rhyzopertha dominica*. Insect Biochem 10:147–150
Price NR, Mills KA, Humphries LA (1982) Phosphine toxicity and catalase activity in susceptible and resistant strains of the lesser grain borer (*Rhyzopertha dominica*). Comp Biochem Physiol 73C:411–413
Sacktor B (1965) Energetics and metabolism of muscular contraction in insect mitochondria. In: Rockstein M (ed) Physiology of insecta, vol II. Academic Press, London New York, pp 484–580
Sacktor B (1974) Biological oxidations and energetics in insect mitochondria. In: Rockstein M (ed) Physiology of insecta, vol IV. Academic Press, London New York, pp 271–353
Sacktor B, Childress CC (1967) Metabolism of proline in insect flight muscles and its significance in stimulating the oxidation of pyruvate. Arch Biochem Biophys 120:583–588
Tischler N (1936) Studies on how derris kills insects. J Econ Entomol 28:215–219
Tsuda S, Urakawa N, Fukami J (1977) Inhibitory effect of papaverine on a respiration-dependent contracture of guinea pig *Taenia coli* in high K^+ medium. III. Difference effect of papaverine and rotenone on DT-diaphorase. Jpn J Pharmacol 27:855–863
Whitehouse HW (1964) Biochem Pharmacol 13:319
Yamaguchi I, Matsumura F, Kadous AA (1979) Inhibition of synaptic ATPase by heptachlor epoxide in rat brain. Pestic Biochem Physiol 11:285–293
Yamaguchi I, Matsumura F, Kadous AA (1980) Heptachlor epoxide: effect on calcium mediated transmitter release from brain synaptosomes in rat. Biochem Pharmacol 29:1815–1823

The Action of Spider Toxins on the Insect Nerve-Muscle System

P. N. R. Usherwood [1]

1 Introduction

In the search for new classes of pesticides it is perhaps surprising that only relatively scant attention has been given to the natural products found in the venoms of a variety of insect predators. Extensive use has been made of vertebrate and invertebrate toxins to gain information on the structure and function of central and peripheral nervous systems across the animal kingdom but for a variety of reasons these are of little interest to the chemical industry. However, many insect predators produce venoms, most of which remain largely uncharacterized, which may well contain active principles of commercial interest. Insect neurobiologists have compelling reasons to ponder over the possible potential of these compounds as research tools since many of the developments which recently have taken place in neuroscience have depended upon the use of venoms and toxins. For example, toxins from snake and spider venoms have been used to study transmitter storage, release and turnover at peripheral and central synapses and the snake toxin, α-bungarotoxin has been particularly instrumental in providing an understanding of the molecular properties of the nicotinic acetylcholine receptor protein.

My laboratory research is directed mainly towards the nerve-muscle junctions which characterize insect skeletal muscles. α-Bungarotoxin has no effect at either the excitatory (glutamatergic) or inhibitory (GABAergic) junctions on these muscles but other toxins, such as *Lactrodectus* toxin from the black widow spider, have identical effects on insect and vertebrate synapses where they block transmission presynaptically (Cull-Candy et al. 1973). Toxins present in the venom of the braconid wasp *Microbracon hebetor* also block transmission presynaptically at insect glutamatergic nerve-muscle junctions (Piek 1966a, 1969) but according to Piek (1984) this venom contains labile toxins of high molecular mass and is, therefore, of limited interest to the pesticide industry. However, other studies in Piek's laboratory have shown that potent toxins of low molecular weight are present in venoms of some parasitic insects (Piek 1966b, Piek and Spanjer 1978).

In this review I will describe some potent blockers of transmission at locust excitatory nerve-muscle junctions which have been isolated from spider venoms and which have molecular masses < 1 K Daltons. The objectives of this work were not primarily to identify candidates for the pesticide industry but rather to obtain toxins from insect predators which could be used to isolate and character-

1 Department of Zoology, Nottingham University, University Park, Nottingham NG7 2RD, United Kingdom

ize the glutamate receptors found postsynaptically at excitatory nerve-muscle junctions on insect skeletal muscle and which play such a vital role in motor control in insects. Glutamate synaptic receptors are present throughout the animal kingdom and have been extensively studied, but investigations of these structures have been impeded by the lack of specific antagonists of the irreversible or slowly-reversible type, such as the nicotinic acetylcholine receptor antagonist α-bungarotoxin (Lee 1970, Barnard 1982). However, studies in my laboratory over the past 3 years and by Tashmukhamedov et al. (1983) suggest that the venoms of araneid spiders (Orb Web spiders) contain one or more components which specifically antagonize glutamate receptor-channel complexes and that these may eventually provide neurobiologists with suitable compounds for the isolation and characterization of glutamate receptors. Kawai and colleagues have already demonstrated specific competitive antagonism of the postjunctional glutamate receptors of lobster glutamatergic nerve-muscle junctions by extracts of venoms from the araneid spiders, *Nephila clavata* and *Araneus ventricosus* (Kawai et al. 1982 a, b). *Nephila* venom suppresses the excitatory (glutamatergic) postsynaptic potentials of lobster muscle, without affecting the inhibitory (GABAergic) nerve-muscle junctions on this muscle, and blocks muscle depolarizations caused by L-glutamate without affecting those caused by L-aspartate (Kawai et al. 1983).

In this chapter I will review the work on spider venoms undertaken in my laboratory and attempt to place it in the context of data which has emanated from other sources during the past few years. Most of my review is based upon information which has already been published (Boden et al. 1984, Usherwood et al. 1984, Bateman et al. 1985).

2 A Preliminary Survey of the Action of Spider Venoms at the Locust Nerve-Muscle Junction

Crude venom or venom gland extracts from species of 9 spider genera have been tested on two nerve-muscle preparations from adult locusts (*Schistocerca gregaria*), the metathoracic retractor unguis (Usherwood and Machili 1968) and extensor tibiae (Hoyle 1955) preparations. Either milked venom or homogenized venom gland was dissolved in locust saline (Clark et al. 1979) to notional concentrations (either milked venom from one spider or one venom gland, diluted in 1 ml of locust saline equals 1 U of concentration). Full experimental details are to be found in Usherwood et al. (1984) and Bateman et al. (1985).

All of the venoms depressed the neurally evoked twitch contraction of the retractor unguis muscle, but unlike the rest those of *Apstopstichus* spp., *Aliatypus* spp. and *Steotoda grossa* had no effect on depolarizations evoked at excitatory junctions of extensor tibiae muscle during glutamate ionophoresis (Table 1). This suggests that the venoms of *Apstopstichus, Aliatypus* and *Steotoda* contain toxins which block transmission presynaptically at locust excitatory nerve-muscle junctions. In this respect, at least superficially, they resemble *Lactrodectus* venom (Cull-Candy et al. 1973). Crude venom or venom gland extracts from *Hololena* spp., *Neoscona arabesca, Araneus gemma* and *Argiope trifasciata* reduced the am-

Table 1. Effect of spider venoms on locust nerve-muscle preparations[a]. (From Usherwood, Duce and Boden, unpublished data)

Species	Effect of milked venom or venom gland extract
Aliatypus spp.	30% depression of twitch contraction of retractor unguis muscle. Rapid recovery. No effect on junctional glutamate potential at 0.01 U
Apstopstichus spp.	90% depression of twitch contraction with 1 U. No recovery. No effect on junctional glutamate potential at 0.5 U
Araneus gemma	0.5 U gives 100% depression of twitch contraction. Slow reversibility – stimulus frequency dependent onset and recovery. Complete loss of junctional glutamate potential at 0.005 U
Argiope trifasciata	As for *Araneus gemma*
Bothriocyrtum californicum *Hebestatis theveneti*	90% depression of twitch concentration of retractor unguis muscle with 0.5 U. Incomplete recovery on removal of venom extract. 60% depression of junctional glutamate potential with 0.01 U
Chiracanthium inclusum	Initial twitch potentiation with 0.5 U then depression of twitch to >80%. No reversibility. With 1 U a large contracture preceded the depression of twitch amplitude. Slight hyperpolarization of muscle with 50% reduction of junctional glutamate potential at 0.5 U
Hololena spp.	50% depression of twitch contraction with 0.5 U. Slight recovery only. 95% reduction of junctional glutamate potential with 0.01 U
Neoscona arabesca	50% depression of twitch contraction with 0.5 U. Rapid recovery on removal of venom gland extract. 80% reduction in junctional glutamate potential at 0.1 U. Slow recovery
Steotoda grossa	100% depression of twitch contraction with 0.3 U. No recovery. No effect on junctional glutamate potential at 0.3 U

[a] 1 U = contents of 1 venom gland in 1 ml locust saline. All experiments undertaken at room temperature (20–25 °C)

plitude of the retractor unguis muscle twitch and the extensor tibiae muscle junctional glutamate potential in a manner consistent with activation-induced postjunctional block. In other words these venoms were effective in blocking the glutamate receptor channel complexes located in the postjunctional membrane of the excitatory nerve-muscle junctions only when the latter interacted with either transmitter or L-glutamate. This suggests that the site of action of the active components in the venom is the open glutamate receptor channel. The enhanced rate of onset and extent of block of nerve-muscle transmission by these venoms observed when either the frequency of the stimulation of the retractor unguis muscle or the glutamate pulse frequency in the ionophoretic experiments was increased, supports this suggestion.

3 Pharmacological Studies of *Araneus* and *Argiope* Venoms

For reasons of availability and because of their high potency it was decided to investigate further the properties of venoms from the Orb Web spiders, *Araneus gemma* and *Argiope trifasciata*. Similar results were obtained with either freshly frozen venom gland extracts, extracts of venom glands from freshly killed spiders or freshly frozen, milked venom. In excess of 200 nerve-muscle preparations have been used so far in these studies. The results obtained with *Argiope* venom are qualitatively similar to those obtained with *Araneus* venom.

3.1 Action of Crude Venom on the Retractor Unguis Muscle

The neurally-evoked twitch contractions of the locust retractor unguis muscle were rapidly abolished by 1 U of *Araneus* or *Argiope* venom (Fig. 1). The rate and extent of the depression of twitch amplitude was dependent upon the stimulation frequency. In the absence of neural stimulation the venom seemingly had little effect on the nerve-muscle preparations, although sometimes a small contracture of the muscle occurred immediately after its application, reflecting perhaps the presence of L-glutamate or its agonists in the venom.

Recovery from the effects of the venom was very slow (Fig. 1) i.e. much slower than expected from simple diffusion of material from extracellular space within the muscle. The rate of recovery was depressed if the muscle was stimulated during the recovery period.

Application of either L-glutamic acid (10^{-4} M) or L-quisqualic acid (5×10^{-4} M) to retractor unguis muscle usually evokes a phasic contraction (Usherwood and Machili 1966, 1968, Clements and May 1974). This was greatly reduced or even abolished in the presence of 1 U of venom, especially if application of agonist was preceeded by neural stimulation of the muscle in the presence of venom.

These data suggest that the venoms of *Araneus* and *Argiope* contain toxins which cause activation-induced block of excitatory nerve-muscle junctions on locust muscle. The slow recovery of these preparations from the effects of the

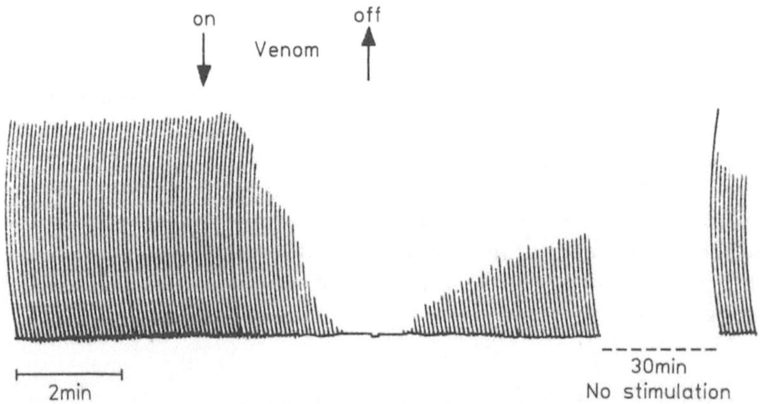

Fig. 1. Abolition and slow recovery of retractor unguis muscle twitch contraction during application of 1 U *Argiope trifasciata* venom gland extract. Note that during the recovery period almost full return of the twitch amplitude occurred initially after a 30-min period during which the preparation was rested, but with repeated stimulation the twitch amplitude fell again. The muscle preparation was stimulated at 0.25 Hz. (From Boden, Duce and Usherwood unpublished data)

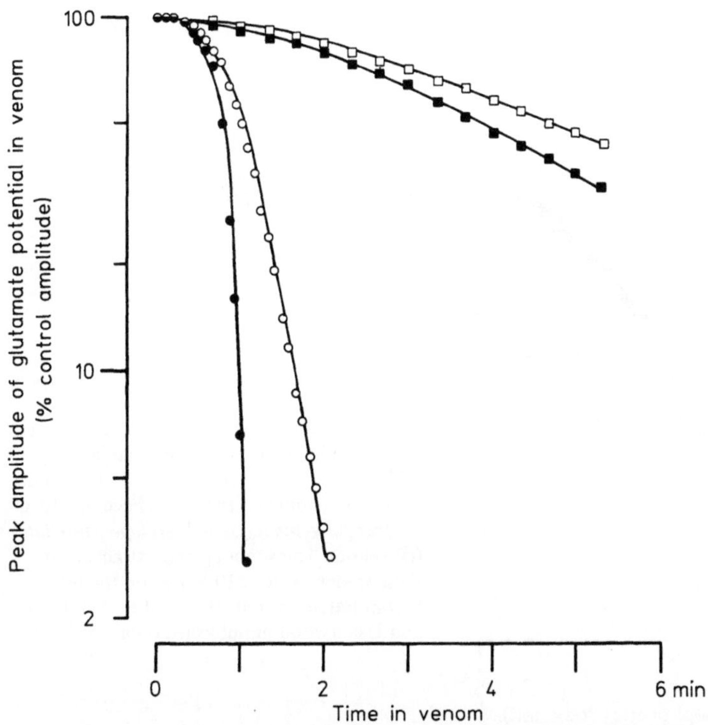

Fig. 2. Semi-logarithmic plots of time course of onset of inhibition of ionophoretic glutamate potentials by < 10 K Dalton fractions of venom gland extracts of *Argiope trifasciata* (●, ■) and *Araneus gemma* (○, □). Notional concentration of venom was 0.1 U (○, ●) and 0.01 U (□, ■). Ionophoretic responses recorded from superficial nerve-muscle junctions on locust extensor tibiae muscle. (From Bateman et al. 1985 and Boden, Duce and Usherwood unpublished data)

venoms indicate that the toxin(s) they contain bind tightly near to the channel of the postjunctional glutamate receptor, from which site they can block the channel when it opens. To test this proposal further experiments were undertaken using ionophoretic and voltage clamp techniques.

3.2 Action of Venom on the Junctional Glutamate Potential of Extensor Tibiae Muscle

Concentrations of venom as low as 0.005 U depressed the postjunctional response of excitatory synapses on extensor tibiae muscle to glutamate ionophoresis (Fig. 2). The rate and extent of this depression was dependent upon the ionophoretic pulse repetition frequency, even after blocking glutamate receptor desensitization with concanavalin A (Mathers and Usherwood 1976,1978). Recovery of the ionophoretic response was either undetectable or incomplete during the limited recording period (ca. 10 min). In some preparations the venom depolarized the muscle by a few mV. Dose-response curves constructed using ionophoretic data

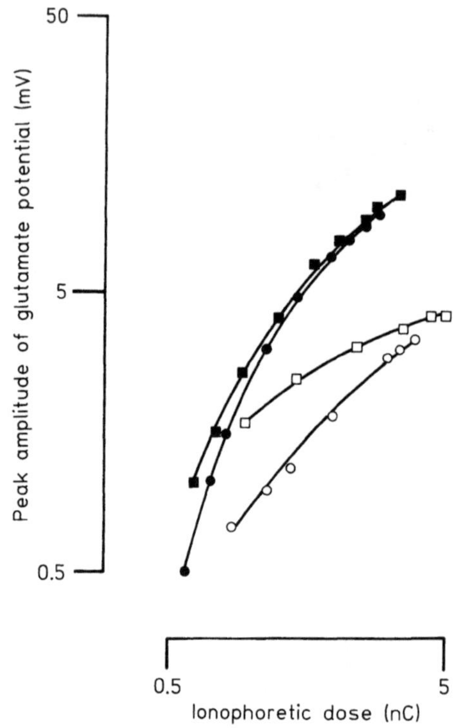

Fig. 3. Glutamate ionophoretic dose-response data recorded from two superficial junctions on locust extensor tibiae muscle in the absence (■, ●) of venom and in the presence of 0.02 U of *Araneus gemma* (○) and *Argiope trifasciata* (□) venom. The venom gland extracts from these spiders were < 10 K Dalton fractions. (From Bateman et al. 1985 and Boden, Duce and Usherwood unpublished data)

were shifted to the right in a non-parallel fashion by the venoms, suggesting uncompetitive antagonism or open channel block (Fig. 3). However, the slow reversibility of the venom complicates interpretation of the dose-response data using progressively increasing doses of agonist, even with 1 min intervals between the doses as in Fig. 3.

3.3 Action of Venom on the Excitatory Postsynaptic Potential (EPSP) and Excitatory Postsynaptic Current (EPSC) Recorded from Extensor Tibiae Muscle

0.5 U of venom completely abolished the intracellularly recorded EPSP only during repetitive stimulation of the extensor tibiae nerve. Indeed the rate and extent of reduction in EPSP amplitude was dependent both upon the concentration of venom applied to the muscle and the frequency of neural stimulation. Recovery from the effects of the venom was slow, but it was enhanced if the preparation was rested.

Both *Araneus* and *Argiope* venoms reduced the amplitude of the EPSC recorded from extensor tibiae muscle fibres under voltage clamp. Once again the magnitude and the rate of reduction in amplitude were both venom concentration and stimulation frequency dependent. The reduction in EPSC amplitude was also voltage dependent, increasing with hyperpolarizing of the muscle, and was accompanied by voltage dependent changes in decay time constant and rise time constant (Bateman et al. 1985). All of these effects are consistent with the proposal for open channel block, the active component(s) in these venoms presumably bearing a net positive charge.

3.4 Identification of Toxins in *Araneus* and *Argiope* Venoms

Crude venom from both spiders lost activity if kept at room temperature overnight or at 37 °C for 3 h, but this could be prevented by heating the venoms in saline to 100 °C for ca. 2 min. These and other observations suggest that these venoms contain protease or peptidase and that the active components or toxins are heat stable proteins on peptides. However, boiled venom filtered at 1 K Daltons retained almost full activity when assayed on the retractor unguis preparation which suggests that the toxins are more likely to be peptides of low molecular weight. At this stage in our investigations we cannot exclude the possibility that high molecular weight components present in these venoms may act at excitable sites distinct from the excitatory nerve-muscle junctions. This possibility remains to be tested.

Samples of partially purified *Argiope* venom gland extracts [filtered at 10 K Daltons after partioning across a water/chloroform interface (most activity occurring in the aqueous phase)] were fractionated by HPLC. Activity on retractor unguis preparations of the type seen with crude venom was found in more than one group of fractions with minor quantitative differences in the pharmacology of the fractions. These and other results suggest that the toxins in *Argiope* venom which act at the locust nerve-muscle junction are polypeptides. Since the fractions in which they are found show a strong absorbance at 277 nm they may contain phenolic or indolic residues.

4 Discussion and Conclusions

In many respects the results reported herein are not unlike those obtained by Piek and colleagues for the action of venom of the digger wasp *Philanthus triangulum* on the locust nerve-muscle system (Piek et al. 1971, Clark et al. 1982). One component of this venom, δ-philanthotoxin, inhibits transmission at the excitatory nerve-muscle junction on locust muscle by blocking the channels gated by the postjunctional glutamate receptors. However, whereas recovery from the effects of δ-philanthotoxin is enhanced by agonist activation of the glutamate receptor channel complexes the recovery from *Araneus* and *Argiope* venoms is depressed under this condition. The presence of L-glutamic acid or its agonists in the venom of these spiders which was suggested in some of the experiments reported herein is of considerable interest since it would assists, at least initially, in generating paralysis of prey by activating postjunctional glutamate receptors.

There are many variations in the specifics of predation among the Orb Web spiders. Most species subdue their prey by first wrapping it with silk before biting it. This is especially true with large prey such as grasshoppers. However, with dangerous prey, e.g. wasps or bombadier beetles or prey that might otherwise readily escape, e.g. moths, the bite may precede the wrap. *Nephila,* the tropical Orb Web spider studied by Kawai and colleagues (Kawai et al. 1982a, b, 1983) attacks all of its prey with a bite first and then performs the wrapping whilst sitting in the hub of its web. In general when more than one prey is captured by an Orb Web spider the additional specimens are also wrapped and bitten but then left as a "reserve" food source. The presence of toxins in spider venom which cause activation-induced paralysis of prey might be useful in restraining captured prey whilst keeping them alive as food reserves.

It is difficult to estimate the potencies of *Araneus* and *Argiope* venoms but the toxin purified from *Argiope* venom and which elutes in the same fraction as a 636 Dalton peptide blocks the ionophoretic glutamate potential of locust extensor tibiae muscle at concentrations as low as 10^{-7} to 10^{-8} M. The relatively small size of the toxin molecule and its high potency encourages us to identify it chemically and to examine the feasibility of its synthesis in the laboratory. The availability of such an active molecule would contribute to the development of our knowledge of the insect glutamatergic synapse and perhaps provide leads for commercial exploitation of this part of the insect excitable system as a target for insect pest control.

Acknowledgements. Much of the work described in this review was undertaken by Philip Boden and Ian Duce, Department of Zoology, University of Nottingham and Andrew Bateman and Ann Dell, Department of Biochemistry, Imperial College, London University.

References

Barnard EA (1982) Isolation of receptors from the central nervous system. In: Neuropharmacology of insects. Ciba Found Symp 88. Pitman, London

Bateman A, Boden P, Dell A, Duce IR, Quicke DLJ, Usherwood PNR (1985) Postsynaptic block of a glutamatergic synapse by low molecular weight fractions of spider venom. Brain Res 339, 237–244

Boden P, Duce IR, Usherwood PNR (1984) Activation-induced postsynaptic block of insect nerve-muscle transmission by the low molecular weight fraction of spider venom. Br J Pharmacol, 82, 221p

Clark RB, Gration KAF, Usherwood PNR (1979) Desensitization of glutamate receptors on innervated and denervated muscle fibres. J Physiol (London) 290:551–568

Clark RB, Donaldson PL, Gration KAF, Lambert JJ, Piek T, Ramsey RL, Spanjer W, Usherwood PNR (1982) Block of locust muscle glutamate receptors by δ-philanthotoxin occurs after receptor activation. Brain Res 241:105–114

Clements AN, May TE (1974) Studies on locust neuromuscular physiology in relation to glutamic acid. J Exp Biol 60:673–705

Cull-Candy SG, Neal H, Usherwood PNR (1973) Action of black widow spider venom on an aminergic synapse. Nature (London) 241:353–354

Hoyle G (1955) Neuromuscular mechanisms of a locust skeletal muscle. Proc R Soc London Ser B 143:346–367

Kawai N, Niwa A, Abe T (1982a) Spider venom contains specific receptor blocker of glutaminergic synapses. Brain Res 247:169–171

Kawai N, Niwa A, Abe T (1982b) Effects of spider toxin on glutaminergic synapses in the mammalian brain. Biomed Res 3:353–355

Kawai N, Niwa A, Abe T (1983) Specific antagonism of the glutamate receptor by an extract from the spider *Araneus ventricosus*. Toxicon 21:438–440

Lee CY (1970) Elapid neurotoxins and their mode of action. Clin Toxicol 3:457–472

Mathers DA, Usherwood PNR (1976) Concanavalin A blocks desensitization of glutamate receptors on insect muscle fibres. Nature (London) 259:409–411

Mathers DA, Usherwood PNR (1978) Effects of concanavalin A on junctional and extrajunctional L-glutamate receptors on locust skeletal muscle fibres. Comp Biochem Physiol 59 C:151–155

Piek T (1966a) Site of action of venom of *Microbracon hebetor* Say (Braconidae, Hymenoptera). J Insect Physiol 12:561–568

Piek T (1966b) Site of action of the venom of the digger wasp *Philanthus triangulum* F on the fast neuromuscular system of the locust. Toxicon 3:191–198

Piek T (1969) Action of the venom of *Microbracon hebetor* Say on the hyperpolarizing potentials in a skeletal muscle of *Philosamia cynthia* Hubn. Comp Gen Pharmac 1:117–120

Piek T (1984) Insect venoms and toxins. In: Kerkut GA, Gilbert LI (eds) Comprehensive insect physiology, biochemistry and pharmacology, vol XI. Pharmacology. Pergamon Press, Oxford New York

Piek T, Spanjer W (1978) Effects and chemical characterization of some paralyzing venoms of solitary wasps. In: Shankland DL, Hollingworth RM, Smyth T (eds) Pesticide and venom neurotoxicity. Plenum Press, New York London

Piek T, Mantel P, Engels E (1971) Neuromuscular block in insects caused by the venom of the digger wasp *Philanthus triangulum* F. Comp Gen Pharmacol 2:317–331

Tashmukhamedov BA, Usmanov PB, Kazakov I, Kalikulov D, Yukelson LY, Atakuziev BU (1983) Effects of different spider venoms on artificial and biological membranes. In: Toxins as tools in neurochemistry. de Gruyther, Berlin New York, pp 312–323

Usherwood PNR, Machili P (1966) Chemical transmission at the insect excitatory neuromuscular synapse. Nature (London) 210:634–636

Usherwood PNR, Machili P (1968) Pharmacological properties of excitatory neuromuscular synapses in the locust. J Exp Biol 49:341–361

Usherwood PNR, Duce IR, Boden P (1984) Slowly-reversible block of glutamate receptor-channels by venoms of the spiders, *Argiope trifasciata* and *Araneus gemma*. J Physiol (Paris) 79:241–245

Interaction of Low-Molecular-Weight Effectors with Components of the Nicotinic Cholinergic Synapse: Triphenylmethylphosphonium and Phencyclidine as Inhibitors of Acetylcholinesterase

J. VERDENHALVEN and F. HUCHO [1]

1 Introduction

Chemical transmission of nerve impulses at synapses consists of a multitude of sequential molecular steps. The whole process can be modified by interfering with any one of these individual steps. In the case of cholinergic transmission the following steps have to be considered (Fig. 1); (1) transmitter synthesis, (2) transmitter packaging in synaptic vesicles, (3) fusion of vesicles with the presynaptic membrane, (4) release of the transmitter into the synaptic cleft and diffusion to the postsynaptic membrane, (5) transmitter binding to postsynaptic receptors and opening of ion channels, (6) inactivation of the transmitter by enzymatic hydrolysis, (7) presynaptic reuptake of the cleavage products (Hucho 1982).

Insecticides aimed at the nervous system have been developed only for few of these steps of synaptic transmission (Corbett et al. 1984). Actually for the cholinergic system it is only step 6 for which compounds – the organophosphates and the carbamates – are used as insecticides on a large scale. Nature has been much more inventive and variable in creating toxins specifically affecting the transmitter pathway. But interestingly there is at present not a single naturally occurring compound known blocking step 6, i.e. the acethylcholinesterase, to a significant extent. Most synaptic toxins interact either with the presynaptic release mechanism (step 3 and 4) or with the postsynaptic receptors (step 5). Most prominent in the former group are botulinum toxin, β-bungarotoxin (release blocking) and latrotoxin (release stimulating). Toxins of the latter type are the competitive cholinergic antagonists D-tubocurarine and the α-neurotoxins from snake venoms, and furthermore the non-competitive antagonists histrionicotoxin from the skin of certain Columbian frogs and phencyclidine (PCP).

While the development of insecticides affecting cholinergic transmission has so far focussed rather narrowly on the acetylcholinesterase – the one synaptic component not being a target of natural neurotoxins – the concentration of the chemist's efforts on the cholinergic synapses strikes the neuroscientist as even more narrow-minded.

There are definitively more neurotransmitters beyond acetylcholine in the insect nervous system some of which are already known (serotonin, dopamin, octopamin) others remain to be discovered (e.g. neuropeptides). Including the biogenic amine transmitters in insecticide research increases the number of possible targets considerably. These transmitters interact with receptors which are much

1 Institut für Biochemie, Freie Universität Berlin, 1000 Berlin 33, FRG

Approaches to New Leads for Insecticides
Edited by von Keyserlingk et al.

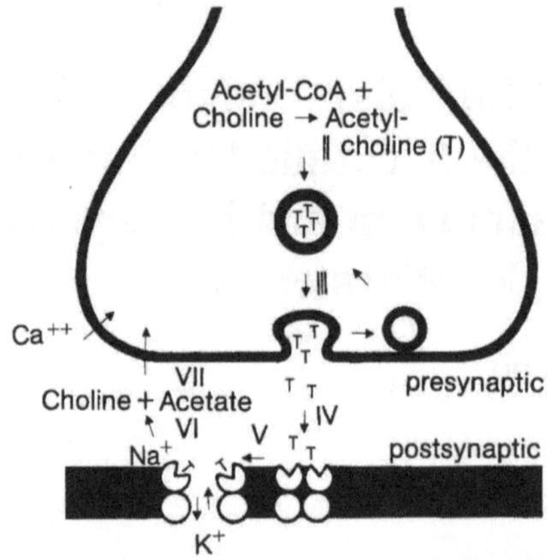

Inhibitor of transmitter inactivation

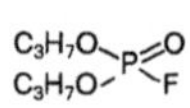

Diisopropylfluoro-phosphate (DFP)

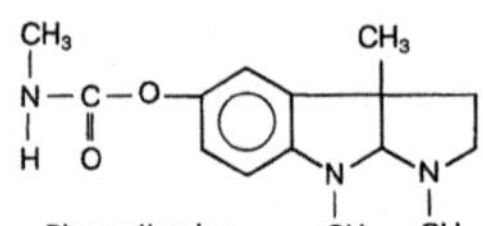

Physostigmine (Eserine)

Presynaptic inhibitor of re-uptake

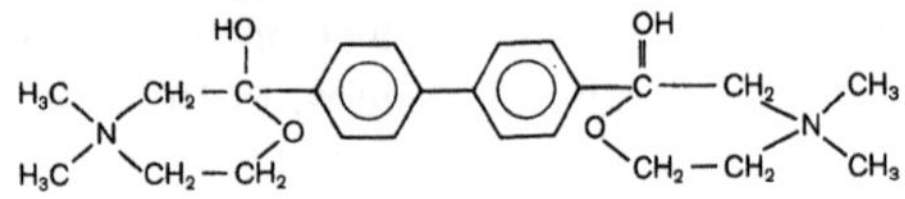

Hemicholinium (HC-3)

Presynaptic blocking agents

β-Bungarotoxin
Botulinum-toxin

Postsynaptic blocking agents

(non competetive)

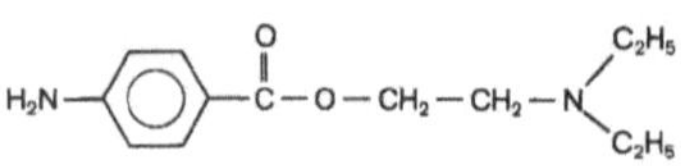

Procaine

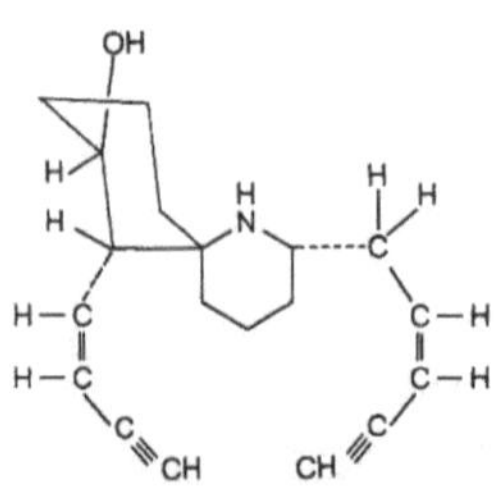

Histrionicotoxin (HTX)

(competetive)

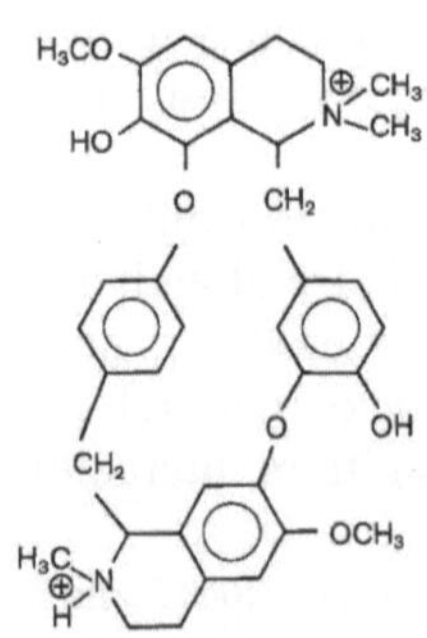

D-tubocurarine

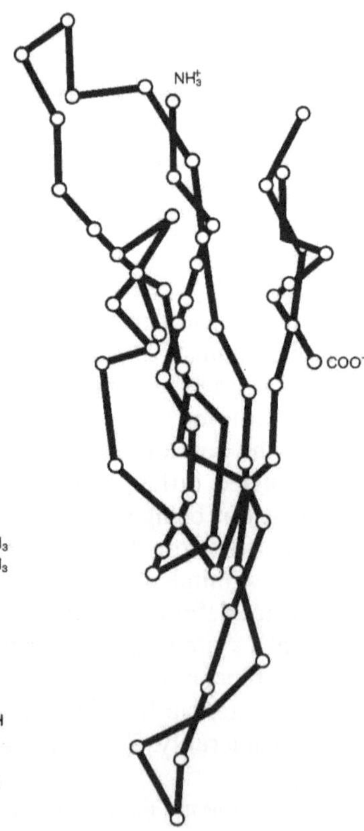

α-neurotoxin from snake venom

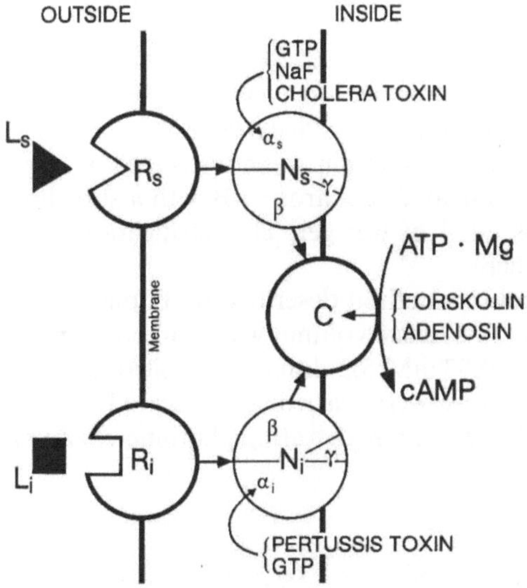

Fig. 2. Schematic representation of signal transduction through a membrane via the second messenger producing enzyme adenylate cyclase. *L* ligand (hormone or neurotransmitter); *R* receptor; *N* coupling protein, composed of three polypeptide chains which are the target of a variety of regulatory effectors and toxins; *C* adenylate cyclase. The index *s* and *i* indicates stimulatory and inhibitory action, respectively (Hildebrandt et al. 1984)

more complicated than the nicotinic cholinergic receptors (Fig. 2). While the latter consist of a binding moiety and an ion channel the former are composed of at least three components (De Lean et al. 1980). Receptor (R), the enzyme adenylate cyclase (C) and a coupling protein (N) which transduces the signal conveyed by the transmitter from R to C (from the extracellular to the intracellular surface of the membrane). The pharmacology of this tripartite system is extensive though little is known about it in insects. Transfer of this pharmacological knowledge to insecticide research should be a challenge for future work.

Even within the compounds used as insecticides inhibiting acetylcholinesterase there has been little variability so far (Fest and Schmidt 1982). They are either directed against the esteratic subsite, an especially reactive -OH group of a serin residue, or against the anionic subsite, carrying a negative charge interacting with the quaternary ammonium group of acetylcholine (Rosenberry 1975). In the present contribution we propose two compounds extending the spectrum of ligands applicable as inhibitors of the active site of the acetylcholinesterase. The one is the psychoactive drug phencyclidine, the other is the triphenylmethylphosphonium cation ($TPMP^+$) which indicates that besides ammonium compounds other positively charged groups such as the phosphonium and possibly arsonium compounds may be candidates for the development of poisons for cholinergic synapses on the basis of acetylcholinesterase inhibition.

Fig. 1 A, B. Schematic representation of the nicotinic cholinergic synapse, including some of the individual steps from which the process of chemical nerve impuls transmission is composed. The figure also contains typical inhibitors of some of these steps

2 Methods

Preparation of acetylcholinesterase: We have purified the collagen tailed form of acetylcholin-esterase from *Torpedo marmorata* by high salt extraction and affinity chromatography on an acridinium-sepharose. The resin was constructed by coupling 9-(aminopropylamino)acridine to activated sepharose 4 B with a substitution grade of 0.2 μM acridine residues ml^{-1} of wet gel. The purification procedure will be published in detail elsewhere.

Acetylcholin-esterase assay: We used the method described by Ellman et al. 1961 with the following modifications. The assay volume was 1 ml containing 1 M NaCl 10 mM Tris · HCl pH 7.4 and 0.33 mM 5,5-dithiobis-(2-nitrobenzoic-acid). The acetylthiocholine concentrations were varied as indicated in the figures. The enzymatic reaction was followed by a registrating photometer (Zeiss DMR 11).

3 Materials

Torpedo marmorata was obtained alive from Prof. Whittaker, Göttingen, triphenylmethylphosphonium bromide from Sigma, München, dithiobis-(2-nitrobenzoicacid) from Serva, Heidelberg, acetylthiocholine jodide, acridone and 1.2 diaminopropane from Fluka, Buchs (Schweiz), CH activated sepharose 4 B from Pharmacia, Freiburg. All other reagents were of the highest purity commercially available.

4 Results and Discussion

The lipophilic cation triphenylmethylphosphonium ($TPMP^+$) has been widely applied as a "membrane potential sensor" accumulating into membrane systems (Vacata et al. 1981, Schuldiner and Kaback 1975) (cells and organelles) in a voltage dependent manner.

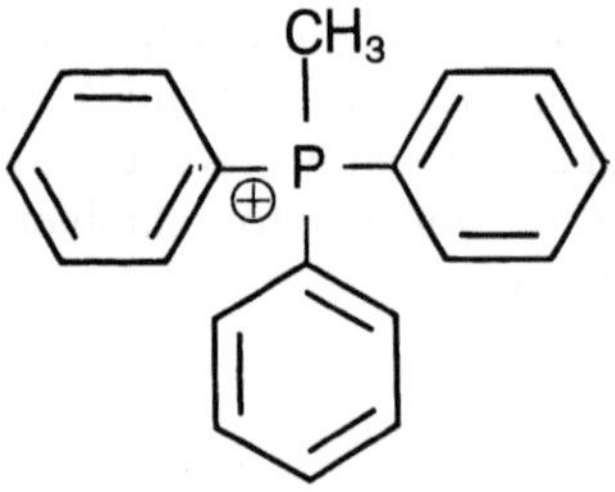

We have found that $TPMP^+$ behaves unusual in this respect when applied to membranes enriched in nicotinic acetylcholine receptor. We observed that in these membranes it accumulates strongly in the presence of a depolarizing cholinergic agonist. This effect was shown to be due to an agonist-stimulated binding

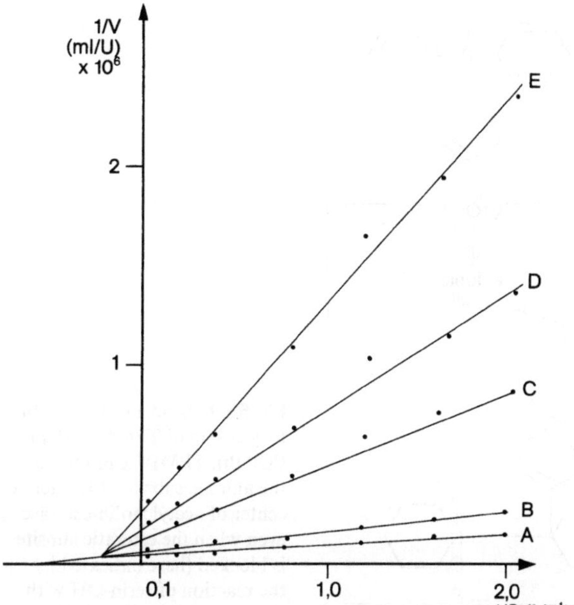

Fig. 3. Lineweaver-Burk plot of the acetylcholinesterase activity in the presence of varying concentrations of $TPMP^+$. Reaction conditions: 1 M NaCl, 10 mM Tris · HCl pH 7.4 Assay according to Ellman with acetylthiocholine as substrate (Ellman et al. 1961), 25 °C. *A* no $TPMP^+$; *B* 50 µM; *C* 125 µM; *D* 250 µM; *E* 500 µM $TPMP^+$

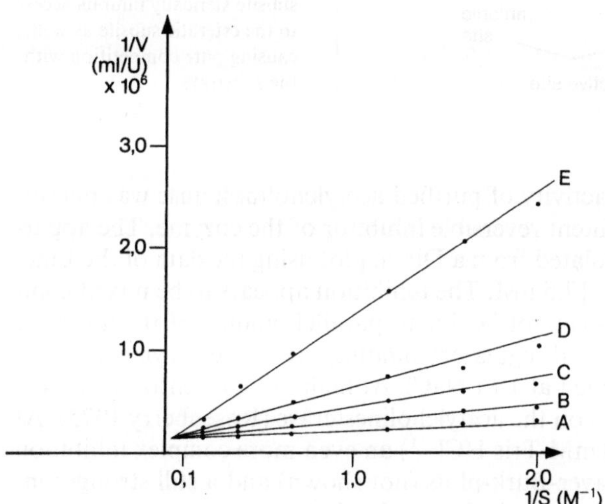

Fig. 4. Lineweaver-Burk-plot of the acetylcholinesterase activity in the presence of varying concentrations of PCP. Conditions as in Fig. 3. *A* no PCP; *B* 50 µM; *C* 125 µM; *D* 250 µM; *E* 500 µM PCP

of [^{3}H]-$TPMP^+$ to the nicotinic acetylcholine receptor (AChR). A more detailed analysis of its binding properties defined [^{3}H]-$TPMP^+$ as a noncompetitive antagonist interacting with the ion channel of the receptor either directly or via an allosteric site of the receptor protein complex.

With these observations in mind we investigated whether $TPMP^+$ interacts with other components of the cholinergic synapse as well. The phosphonium group is sterically very similar to the corresponding ammonium group. Therefore

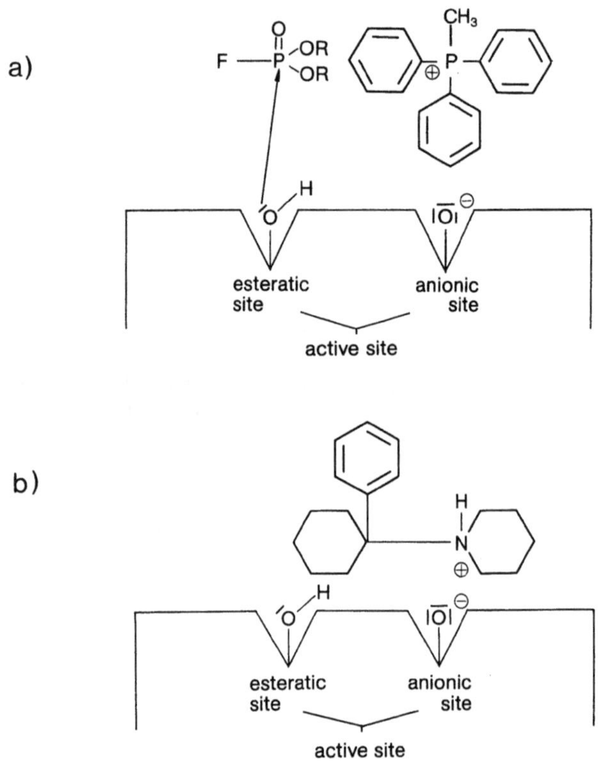

Fig. 5 a, b. Model of the inhibitory action of $TPMP^+$ (**a**) and PCP (**b**). $TPMP^+$ can bind to the anionic subsite of the active center of acetylcholin-esterase even when the esteratic subsite is blocked (here indicated by the reaction of serin-OH with the organophosphate DFP). Binding of PCP to the anionic subsite sterically inhibits access to the esteratic subsite as well, causing pure competition with the substrate

the effect of $TPMP^+$ on the activity of purified acetylcholinesterase was investigated (Fig. 3). $TPMP^+$ is a potent reversible inhibitor of the enzyme. The apparent inhibitor constant as calculated from a Dixon plot using the data of the Lineweaver-Burk-plot, Fig. 3, was 17.5 μM. The inhibition appears to be mixed competitive non-competitive.This cannot be due to parallel binding of the inhibitor to the active site and a peripheral regulatory binding site of the enzyme, because the enzyme assay was performed at 1 M NaCl. At high salt concentrations there are no peripheral binding sites on the acetylcholinesterase (Rosenberry 1975). At lower salt concentrations (10 mM Tris HCl^{-1}) an even more complex inhibition is apparent from the Lineweaver-Burk-plots (not shown) and a still stronger inhibition is observed ($K_i = 0.5$ μM) which may be due to a concerted action of $TPMP^+$ at the active and peripheral binding sites.

The mixed inhibition at high salt concentrations can be explained as follows. The active site for the acetylcholinesterase consists of two subsites, the esteratic site with the activated serin -OH group and the anionic site. The positively charged $TPMP^+$ most likely interacts with the latter. It can do so even after the substrate has been cleaved and the serin -OH has been acylated. The deacylation step is inhibited by $TPMP^+$ bound to the anionic site. This is a classical uncompetitive inhibition (Krupka 1964) overlapping kinetically with the competitive inhibition of the substrate binding step by $TPMP^+$.

References

Corbett JR, Wright K, Baillie AC (1984) The biochemical mode of action of pesticides, 2nd edn. Academic Press, London New York

Ellman GL, Courtney KP, Andres V, Featherstone RM (1961) A new and rapid colorimetric determination of acetylcholinesterase activity. Biochem Pharmacol 7:88–95

Fest C, Schmidt J (1982) The chemistry of organophosphorous pesticides, 2nd edn. Springer, Berlin Heidelberg New York

Hildebrandt JD, Codina J, Risinger R, Birnbaumer L (1984) Identification of a γ-subunit associated with the adenylyl cyclase regulatory proteins n_s and n_i. J Biol Chem 259:2039–2043

Hucho F (1982) Einführung in die Neurochemie. Verlag Chemie, Weinheim Deerfield Beach Basel

Krupka RM (1964) Acetylcholinesterase: Trimethylammonium-ion inhibitor of demethylation. Biochemistry 3:1749–1755

Lauffer L, Hucho F (1982) Triphenylmethylphosphonium as an ion channel ligand of the nicotinic acetylcholine receptor. Proc Natl Acad Sci USA 79:2406–2409

Lean A De, Stadel JM, Lefkowitz RJ (1980) A ternary complex model explains the agonist-specific binding properties of the adenylat cyclase-coupled β-adrenergic receptor. J Biol Chem 255:7108–7117

Rosenberry TC (1975) Acetylcholinesterase. Adv Enzymol 43:103–248

Schuldiner S, Kaback R (1975) Membrane potential and active transport in membrane vesicles from *escherichia coli*. Biochemistry 14:5451–5460

Vacata V, Kotyk A, Sigler K (1981) Membrane potential in yeast cells measured by direct and indirect methods. Biochem Biophys Acta 643:265–268

Vincent J-P, Vignon J, Kartalovski B, Lazdunski M (1981) In PCP. Receptor sites for phencyclidine in mammalian brain and peripheral organs. In: Domino EF (ed) Historical and current perspectives. NPP Books, Ann. Arbor

Neurochemistry of Cholinergic Synapses in Insects

H. BREER [1]

1 Introduction

Important efforts to control insects that are agricultural pests or vectors of infectious and parasitic diseases employ chemicals that attack the nervous system, and in many cases obviously synapses are the sites of action (Brooks 1980). Synapses are the regions where neural communication takes place and furthermore they are the sites where the electrical and chemical activity of neurons are inexorably interlinked and are clearly open to pharmacological interruption in a variety of ways.

If this approach to insect control is to be advanced, it will require a more detailed understanding of the molecular details of synaptic transmission at synapses in the CNS of insects. This includes the presynaptic mechanism just as the processes in the synaptic cleft and the molecular actions of synaptic transmitters at the postsynaptic membrane. Neurochemical studies for these particular problems are almost totally missing for insects.

Table 1. Acetylcholine-content in the CNS of various species

Species	nmol mg^{-1} protein	μg g^{-1} weight	Reference
Fly (Musca)	9.0	150	Lewis (1956)
Cockroach	8.5	136	Smallman et al. (1958)
Locust	7.6	111	Breer (1981 a)
Guinea pig	0.3	4.8	Barker et al. (1972)
Rat	0.2	3.4	Saelens et al. (1973)

Comparing the concentrations of various classical neurotransmitters, e.g. in the CNS of locust, revealed acetylcholine as by far the most frequent transmitters (Breer 1981 a); and a comparison of ACh-concentrations in the CNS of various insects and vertebrates (Table 1) emphasizes the enormous importance of cholinergic innervation in insects. One of the reasons for the preponderance of cholinergic nerve contacts in the ganglia of arthropods is obviously the basically different innervation pattern in these organisms (Florey and Michelson 1974) in particular the cholinergic nature of mechano- and chemosensoric afferences. More detailed neurochemical studies on functional elements of cholinergic synapses seem therefore most urgently needed for insect neuropharmacology.

1 University Osnabrück, Department of Zoophysiology, 4500 Osnabrück, FRG

Approaches to New Leads for Insecticides
Edited by von Keyserlingk et al.

2 Analysis of the Cholinergic Systems

In cholinergic neurons acetylcholine is synthesized from acetyl-CoA and choline by means of the enzyme choline acetyl transferase (ChAT) and is translocated into synaptic vesicles. Depolarization of the presynaptic membrane induces the release of acetylcholine (ACh). The released transmitter is very transiently bound to highly specific receptors and subsequently inactivated by the hydrolytic enzyme acetylcholine-esterase (AChE). The choline is taken up into the presynaptic terminal and is preferentially used for synthesis of ACh.

A comparison of the ACh-concentration as well as the enzymatic parameters, the ACh-synthesizing enzyme ChAT and the inactivating AChE, in the central nervous tissue of locust and mouse (Fig. 1) emphasizes the particular importance of cholinergic innervation for insects.

Neurochemical analysis of further essential activities of the cholinergic system, especially the dynamic processes, like Ch-uptake or transmitter release, require as an essential prerequisite the isolation of intact nerve endings, of synaptosomes.

The discovery, that during homogenization of nervous tissue under appropriate conditions, nerve endings are detached from their axons, reseal and can then be isolated as subcellular structures, as synaptosomes (Whittaker 1965), is now being considered as a landmark in neurochemistry and has opened the possibility to study various aspects of synaptic transmission in the test tube with classical biochemical techniques. Neurochemical research on synaptosomes tries to receive information which help to bridge the gap between the synaptic ultrastructure as seen in the EM and the functional activities of synapses as registered with electrophysiological techniques. Another major advantage of synaptosomes would be the possibility to employ toxins and drugs without the problems of access and the influence of surrounding structures.

All attempts to isolate nerve endings from arthropods using standard techniques have failed; therefore we have developed a microscale floatation procedure

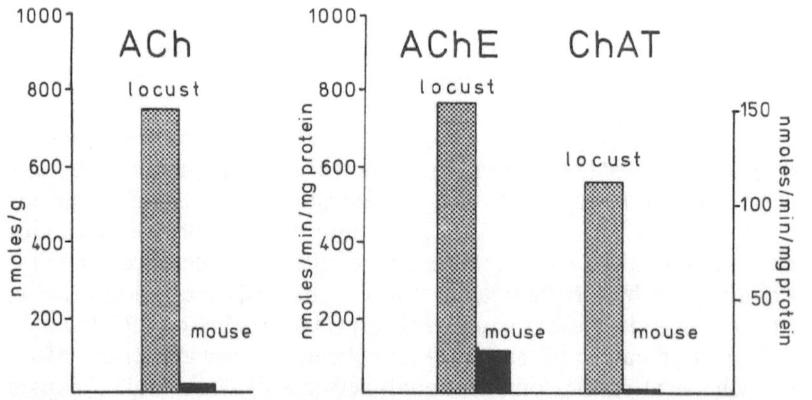

Fig. 1. Concentration and specific activities of cholinergic parameters acetylcholine (*ACh*), acetylcholinesterase (*AChE*), choline acetyltransferase (*ChAT*) in the central nervous tissue of locust and mouse

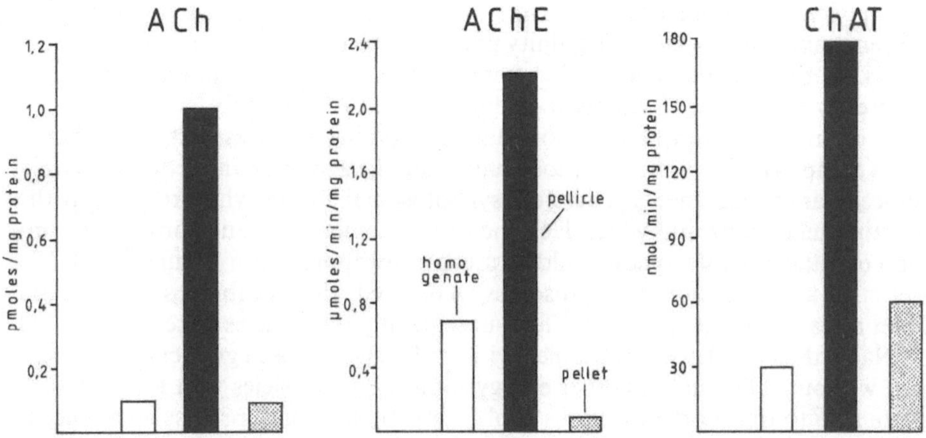

Fig. 2. Distribution of cholinergic activities in the homogenate as well as in the pellet and pellicle fraction after subcellular fractionation of locust nervous tissue using the microscale flotation procedure

for subcellular fractionation of nervous tissue specifically designed for use with insects (Breer and Jeserich 1980, Breer 1981 b). Briefly, ganglia were chopped and gently homogenized in isotonic solution with a microhomogenizer. The P_2-fraction produced by differential centrifugation was resuspended in isotonic FICOLL, during centrifugation of this mixture the particulate material accumulated on the bottom as well as on top of the gradient. Analysis of marker enzymes showed that mitochondrial enzymes like SDH are concentrated in the pellet fraction, whereas synaptosomal markers were enriched in the pellicle fraction. This was confirmed by electronmicroscopy showing that the pellicle consisted of numerous synaptosomal profiles. The isolated nerve endings retained most of the original morphology; they are well-sealed structures containing spherical vesicles, mostly of the electron-lucent type (Breer 1981 b).

Furthermore this preparation – the pellicle fraction – was greatly enriched in cholinergic markers – ACh, ChAT, AChE – (Fig. 2) indicating that the insect synaptosomal fraction contained high proportions of cholinergic nerve endings, which can be used to study neurochemical aspects of cholinergic nerve terminals in vitro.

2.1 Choline Uptake

Isolated nerve terminals from locusts accumulate exogenous choline in a linear, temperature-sensitive process which is abolished by detergents. Concentration dependent experiments showed saturation indicating that choline is taken up in a carrier-mediated process and analysis of the transport kinetics in a Hofsted plot revealed the existence of a low and a high affinity system, with K_T of 1 µM (Breer 1982). A high affinity choline-uptake system is supposed to be unique to cholinergic nerve endings and to be the rate limiting, regulatory step for the synthesis of acetylcholine (Kuhar and Murrin 1978); and in fact we found that most of the choline taken up via the high affinity pathway was converted to acetylcholine.

What may be the main driving force for this important transport system? Is it directly coupled to an energy transducer such as ATPase?

The interpretation of transport data concerning these aspects have always been complicated by intrasynaptosomal metabolic activities and sequestration in subcellular compartments. Therefore synaptosomal ghosts, which retain only the limiting membrane and which lack the complex energetics and compartimentation of intact synaptosomes would be much more appropriate (Kanner 1983). To get such a preparation, synaptosomes were lysed, the membranes washed and then resealed in isotonic K^+-buffer; subsequently the vesicles were resuspended in NaCl; thus vesicles were obtained with artificially imposed gradients: Na_o^+ K_i^+, but without ATP and any other energy source. Such vesicles with K_i^+ and Na_o^+ were able to transport exogenous choline in a high affinity process and concentrate choline up to a factor of 100 (Breer 1983a). This enormous concentrative effect was energized by ion gradients, only. Therefore, a distinct manipulation of ionic conditions should give some new insight in the functional importance of intra- and extracellular ions for the translocation of choline. Substitution of Na^+-ions caused always a blockade of choline uptake; Na^+ could not be replaced by any other monovalent ions. Vesicles which contained Na^+ inside as well as outside did not transport any choline; indicating that a gradient of Na^+ (o>i) is obviously needed. This result is supported by effects of specific ionophores, as e.g. Monensin, which destroys Na^+-gradients and blocks choline uptake (Fig. 3). Therefore the Na^+-gradient must be considered as the main driving force for the high affinity choline transport. As a consequence it should be expected that the

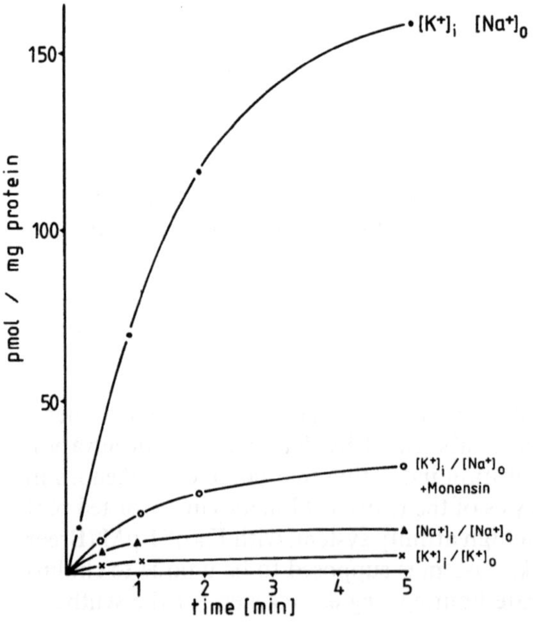

Fig. 3. High affinity choline uptake by membrane vesicles prepared from locust synaptosomes: importance of extracellular ions and sodium gradients for choline influx is shown

transport rate is proportional to the respective Na^+-gradient. Therefore we have analysed the choline transport over a range of Na^+-gradients. In such experiments we found a sigmoidal relationship between choline accumulation and the Na^+-gradient suggesting that more than one Na^+ interacted with the Ch-transport system. When these data were analysed in a Hill plot, the slope of the line was about 2, suggesting a stoichiometry of 2 Na^+ per molecule of choline transported (Breer and Lueken 1983). In similar experiments for the anion specificity it was shown, that a chloride gradient is important as well; however only 1 Cl^- seems to be involved in Ch-transportation. Thus obviously a complex of 2 Na^+ : 1 Cl : 1 Ch is translocated. This points to an electrogenic transport process, that means net charges cross the membrane. If this is the case, the membrane potential which exists as charge difference between inside and outside, should affect the choline-transport; experiments using pharmacological compounds which modify the membrane potential confirm this thesis: Valinomycin, a K^+-ionophore which increases the membrane potential, stimulated the Ch-uptake; CCCP, a H^+-ionophore which destroys the membrane potential, reduced the Ch-uptake.

The results clearly demonstrate that synaptosomal membranes from insects do contain high affinity choline carriers, which are mainly driven by ion gradients.

Accordingly the transport process is probably regulated by modulating the ion gradients. This is confirmed in experiments with various neurotoxins; the uptake of choline into membrane vesicles was significantly inhibited by veratridine, a plant alkaloid, which activates Na^+-channels (Narahashi 1974). In addition it is shown, that tetrodotoxin, a specific blocker of Na^+-channels, which by itself has not much of an effect, almost completely prevented the inhibitory effect of veratridine (Breer and Lueken 1983). To further extent this aspect, the influence of toxins from the venom of the scorpion *Tityus serrulatus* was studied; this venom is known to contain components which cause a delayed inactivation of Na^+-channels (Cahalan 1974). These toxins are obviously active on insect preparations as well, they potentiate the effect of veratridine on choline uptake; the combined actions of veratridine and scorpion venom caused a drastically reduced rate of choline accumulation (Breer 1983 b). However the choline transport may also be regulated by other mechanisms as well. Very recently we found that ATP might be a modulator, an inhibitor of the choline-uptake system: extracellular ATP inhibits Ch-transport in micromolar concentrations. This finding is of particular interest, since ATP might be released together with ACh at cholinergic synapses in insects, as has already been shown for other cholinergic nerve endings (Meunier et al. 1975). The mode of action how ATP affects the choline carrier (binding, phosphorylation) remains to be clarified.

2.2 Transmitter Release

Synaptosomal preparations from insects have the enzyme activity to synthesize ACh and organells for its storage. The release of ACh from insect synaptosomes was studied using a small perfusion technique which prevents re-uptake and allows a continous measurement of the released material. Synaptosomes were im-

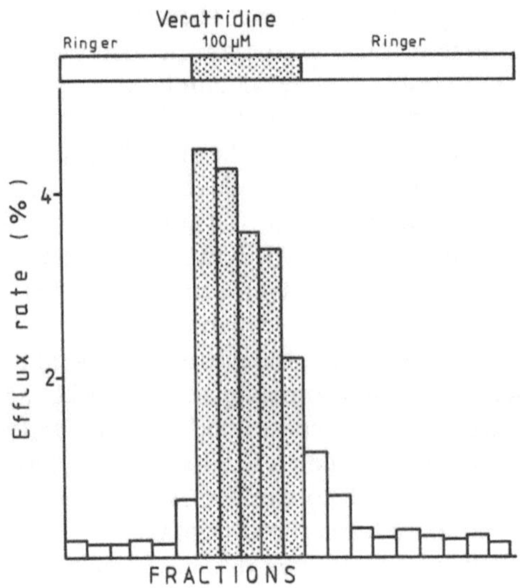

Fig. 4. Release of acetylcholine from insect synaptosomes induced by veratridine. Synaptosomes, preincubated with tritiated choline and immobilized on a small column are perfused with isotonic media containing depolarizing concentration of veratridine; small fractions of the perfusate were collected and analysed for radioactivity and acetylcholine

mobilized on small columns and perfused with isotonic media; samples of the perfusate were collected at various time intervals (Breer and Knipper 1985). Perfusion with isotonic media caused only little efflux; however the addition of depolarizing agents, like micromolar concentrations of veratridine induced a considerable release of material. After the initial peak the release rate decreased and returned to the baseline (Fig. 4). The released compounds were identified as mostly acetylcholine.

The evoked release of ACh from insect nerve terminals was demonstrated to be essentially dependent on extracellular Ca^{2+}. Perfusion with calcium-free medium containing depolarizing levels of K^+ caused no efflux. Only the addition of Ca^{2+} induced the release of ACh. The effect of Ca^{2+} is concentration dependent reaching an optimum at 4 mM. When Ca^{2+} was replaced by any other divalent ions, no release was achieved (Breer and Knipper 1984). These experiments show, that insect synaptosomes provide a valuable system to simulate physiological transmitter release in vitro. Release experiments in the presence of cholinergic ligands have shown that muscarinic agonists, like oxotremorine, significantly reduced the extent of ACh release, whereas antagonists like atropine prevent this depressant effect. In contrast, nicotinic ligands did not show any significant effect at all. The results suggest that on cholinergic nerve terminals of insects the transmitter efflux is regulated by negative feedback mechanisms, probably via muscarinic autoreceptors (Breer and Knipper 1984).

2.3 Acetylcholine Receptors

Receptors for ACh are of particular functional importance in the nervous system of insects as recognition molecules and as transducer of molecular signals. From vertebrate neuropharmacology it is known that there are two types of receptors

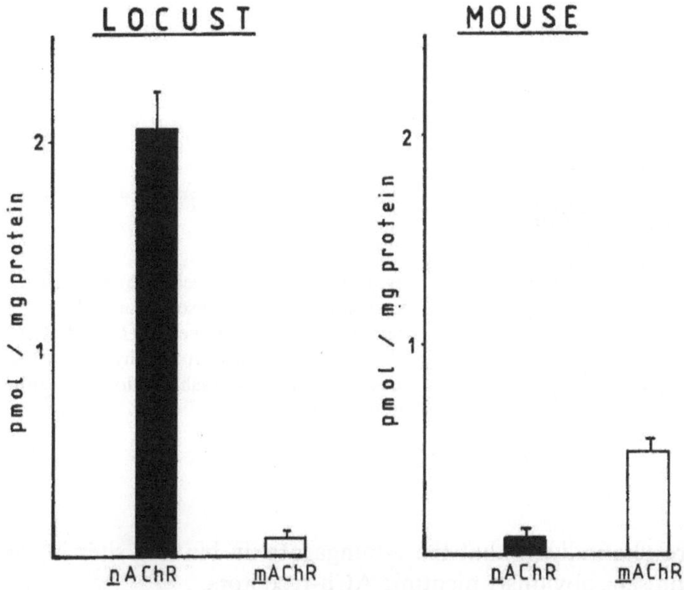

Fig. 5. Comparison of putative nicotinic and muscarinic receptors for acetylcholine in the central nervous tissue of insects. Note the preponderance of nicotinic binding sites in the insect, whereas the muscarinic type predominates in the mouse

for ACh, which differ in pharmacology and mode of action: nicotinic and muscarinic receptors. Both receptor types can be identified and analyzed by means of radioactive-labelled ligands: labelled α-bungarotoxin (BGT), a polypeptid from snake venom, which blocks neuromuscular transmission by interacting with the ACh-receptors is used as a ligand for the nicotinic receptor, whereas tritiated QNB is used for the muscarinic receptor type.

Comparing the binding capacity for these two ligands in the central nervous tissue shows that the receptor concentration in insects is much higher than in mammals (Fig. 5); furthermore it is most remarkable that a different receptor type predominates: in mouse brain mostly QNB-binding sites, putative muscarinic receptors; in locust ganglia very high concentrations of α-toxin binding sites, putative nicotinic receptors. The observation, that insects as well as mammals, which are far apart on the evolutionary scale, do contain nicotinic and muscarinic binding sites, shows that the classical dichotomy of cholinergic receptors developed very early in evolution.

Further characterization of the α-toxin binding in insects revealed that the binding isotherm for toxin shows saturation in the nanomolar range, indicating that there is a limited number of specific binding sites, 2,000 fmol mg^{-1} protein which bind the ligand with very high affinity (K_D 10^{-9} M). The toxin binding sites show distinct nicotinic pharmacology (Breer 1981c). Furthermore in electrophysiological studies Dr. Sattelle has recently clearly demonstrated that at cholinergic synapses in insects, the excitatory postsynaptic potential, the receptor-mediated reaction of ACh, is blocked by α-BGT in nanomolar concentrations (Sattelle

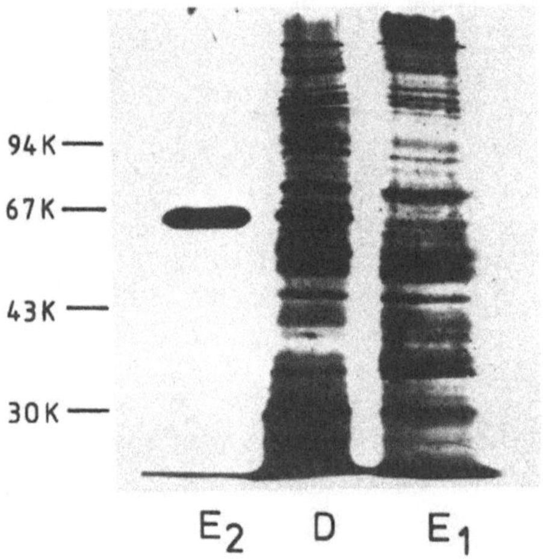

Fig. 6. Polyacrylamide gel electrophoresis of polypeptides from an insect neuronal membrane preparation before (*D*) and after (E_1, E_2) affinity chromatography on α-toxin Sepharose 4 B. E_1 = wash fraction; E_2 = tubocurarineluate

et al. 1983). These results indicate that the α-bungarotoxin binding sites in the ganglia of insects represent obviously nicotinic ACh-receptors.

To compare some molecular aspects of the toxin binding sites from insects with the well characterized ACh-receptor from electric tissue, the cross-reactivity of monoclonal antibodies against the Torpedo receptor with membrane preparation from locust were analysed (Fels et al. 1983). In series of ELISA-tests it was achieved that some of the Torpedo antibodies showed a strong cross reactivity with insect membranes; pointing to some similarities in the molecular structure of the BGT-binding components in the Torpedo electroplaques and the ganglia of locusts.

For further biochemical characterization the receptor was solubilized using mild detergent, which allows to preserve the toxin binding and its pharmacology. Analysis of the binding sites in analytical linear sucrose density gradients revealed that the binding components sedimented as a single species between the two markers: catalase and phosphatase. This corresponds to a sedimentation coefficient of 9 S.

To further purify the receptor, the affinity chromatography technique was employed. Affinity gel was prepared by covalently coupling α-BGT on activated Sepharose 4 B. The gel binds the ACh-receptors whereas all other proteins remain in solution and can be washed out with buffer. The bound receptor proteins were competetively eluted with cholinergic ligands: thus an enormous separation effect was achieved. The protein pattern of the two eluates was analysed using a SDS polyacrylamide gel electrophoresis (Breer et al. 1984). In the fraction that was adsorbed to the Sepharose gel and eluted with curare, there was only one major band corresponding to an apparent molecular weight of about 65,000 (Fig. 6).

Further evidence that the isolated 65 K polypeptide is in fact a constituent of the acetylcholine receptor was achieved by an immunoreaction with a monoclonal antibody which is known to interact with the ligand binding site of the receptor from Torpedo (Watters and Maelicke 1983) and from locust (Fels et al. 1983) using the immunoblotting technique (Burnette 1981) (Fig. 7b). This emphasizes

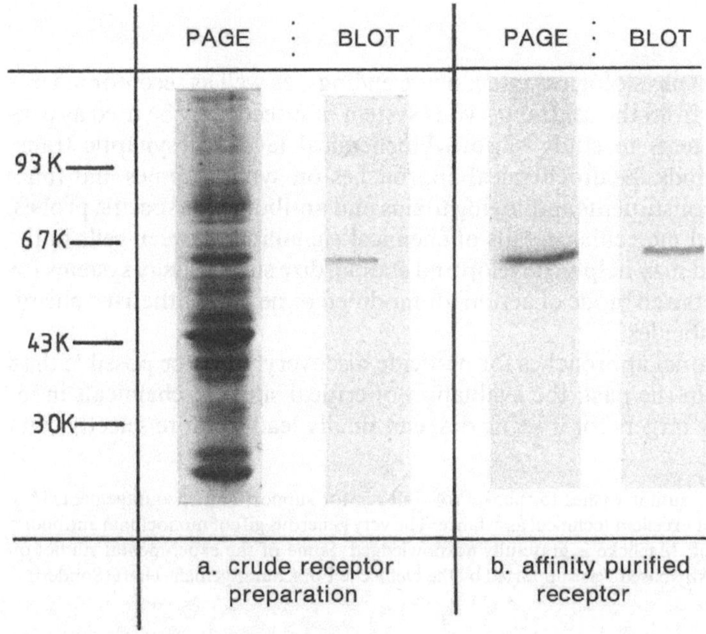

Fig. 7A, B. Immunochemical analysis of acetylcholine receptor protein from locust: **A** Polyacrylamide gel electrophoresis of a crude receptor preparation and an immunoblot with antiserum raised in rabbits against the 65 K protein. The antibodies selectively recognized the receptor polypeptid. **B** Polyacrylamide gel electrophoresis of the affinity purified receptor protein and an immunoblot with a monoclonal antibody against the Torpedo receptor. The pronounced crossreactivity indicates some molecular similarities between both receptor types

the point that the 65 K-polypeptide is a binding side for cholinergic ligands, probably a subunit of the receptor for acetylcholine in the central nervous system of insects. This specific transmitter receptor is probably an oligomer of 65 K molecular weight subunits and is therefore significantly different from the ACh-receptor of electric fish. Meanwhile antisera have been raised against the 65 K receptor component and the immunoblot with a partially purified membrane preparation shows that the antiserum is monospecific: it selectively recognized the 65 K band (Fig. 7a). This antiserum can now be used as a tool in immunochemical and immunocytochemical analysis to evaluate the molecular organization and cellular localization of the ACh-receptors in insect central nervous systems.

In very recent experiments we have obtained significant indications that the receptor is obviously a phosphorylated protein: incubation of purified receptor protein with γ^{32}P-ATP and an endogenous membrane-bound kinase caused a significant phosphatase-sensitive labelling of the receptor. Various molecular aspects of the phosphorylation/dephosphorylation cycle as well as functional implications of the receptor phosphorylation, especially concerning modulation of membrane function, are the subject of ongoing studies. These catalized reactions may be very sensitive switches for changing membrane and probably neuronal functions.

3 Conclusion

Morphological and physiological intact nerve endings, as well as receptor macromolecules isolated from the central nervous system of insects may be used as convenient model systems to study various biochemical facets of synaptic transmission in arthropods. Neurochemical approaches on synaptosomes and functional membrane constituents employing toxins and antibodies as specific probes, will help to unravel molecular details of chemical signalling between cells in the nervous system and may help to develop and standardize simple assay systems for screening the targets and mode of action for modified or newly synthesized chemicals, putative insecticides.

Thus more rational approaches for pesticide discovery might be possible than have been typical in the past; the evaluation of critical sites for chemicals in insects, putative new targets for insecticides, can finally lead to more selective and safer compounds.

Acknowledgements. The author wishes to thank Prof. Lueken for support and encouragement; Mrs. Düvel and Miss Hinz for excellent technical assistance. The very generous gift of monoclonal antibodies kindly provided by Prof. Maelicke is gratefully acknowledged. Some of the experimental studies on which this review is mostly based was supported by the Deutsche Forschungsgemeinschaft (Sonderforschungsbereich 171/84).

References

Barker LA, Dowdall MJ, Whittaker VP (1972) Choline metabolism in the cerebral cortex of guinea pigs. Biochem J 130:1063–1080

Breer H (1981 a) Comparative studies on cholinergic activities in the central nervous system of *Locusta migratoria*. J Comp Physiol 141:271–275

Breer H (1981 b) Characterization of synaptosomes from the central nervous system of insects. Neurochem Int 3:155–163

Breer H (1981 c) Properties of putative nicotinic and muscarinic cholinergic receptors in the nervous system of *Locusta migratoria*. Neurochem Int 3:43–52

Breer H (1982) Uptake of (N-Me-^{3}H)-choline by synaptosomes from the central nervous system of *Locusta migratoria*. J Neurobiol 13:107–117

Breer H (1983 a) Choline transport by synaptosomal membrane vesicles isolated from insect nervous tissue. FEBS Lett 153:345–348

Breer H (1983 b) Venoms and toxins in neurochemical research of insects. In: Hucho F (ed) Toxins as tools in neurochemistry. de Gruyter, Berlin, pp 115–125

Breer H, Jeserich G (1980) A microscale flotation technique for the isolation of synaptosomes from nervous tissue of *Locusta migratoria*. Insect Biochem 10:457–463

Breer H, Knipper M (1984) Characterization of acetylcholine release from insect synaptosomes. Insect Biochem 14:337–344

Breer H, Knipper M (1985) Synaptosomes and neuronal membranes. In: Breer H, Miller TA (eds) Neurochemical techniques in insect research. Springer, Berlin Heidelberg New York

Breer H, Lueken W (1983) Transport of choline by membrane vesicles prepared from synaptosomes of insect nervous tissue. Neurochem Int 5:713–720

Breer H, Kleene R, Benke D (1984) Isolation of a putative nicotinic acetylcholine receptor from the central nervous system of *Locusta migratoria*. Neurosci Lett 46:323–328

Brooks GT (1980) Biochemical targets and insecticide action. In: Insect neurobiology and pesticide action. Soc Chem Ind, London, pp 41–55

Burnette WN (1981) Western blotting: electrophoretic transfer of proteins from SDS-polyacrylamide gels to unmodified nitrocellulose and radiographic detection with antibody and radioiodinated protein. Anal Biochem 112:195–203

Cahalan M (1974) Modification of sodium channel gating by scorpion venom. J Physiol (London) 244:511–534
Fels G, Breer H, Maelicke A (1983) Are there nicotinic acetylcholine receptors in invertebrate ganglionic tissue. In: Hucho F, Ovchinicov (eds) Toxins as tools in neurochemistry. de Gruyter, Berlin, pp 127–140
Florey E, Michelson MJ (1974) Occurrence, pharmacology and significance of cholinergic mechanisms in the animal kingdom. In: Michelson MJ (ed) Comparative pharmacology. Pergamon Press, pp 11–40
Hucho F (1981) The nicotinic acetylcholine receptor. Trends Biochem Sci 6:242–244
Kanner B (1983) Bioenergetics of neurotransmitter transport. Biochim Biophys Acta 726:293–316
Kuhar MJ, Murrin LC (1978) Sodium-dependent high affinity choline uptake. J Neurochem 30:15–21
Lewis PR (1956) The free amino acids in invertebrate nerve. Biochem J 52:330–338
Meunier FM, Israel M, Lesbat B (1975) Release of ATP from stimulated nerve electroplaque junctions. Nature (London) 257:407–408
Narahashi T (1974) Chemicals as tools in the study of excitable membranes. Physiol Rev 54:813–889
Osborne MP (1980) The insect synapse: structural and functional aspects in relation to insecticidal action. In: Insect neurobiology and pesticide action. Soc Chem Ind, London, pp 29–40
Saelens JK, Simke JP, Allen MP, Conroy CA (1973) Some dynamics of choline and acetylcholine metabolism in rat brain. Arch Int Pharmacodyn Ther 209:250–258
Sattelle DB, Harrow JD, Hue B, Pelhate M, Gepner JI, Hall LM (1983) α-Bungarotoxin blocks excitatory synaptic transmission between cercal sensory neurones and giant interneurones of the cockroach, *Periplaneta americana*. J Exp Biol 107:473–489
Watters D, Maelicke A (1983) Organization of ligand binding sites at the acetylcholine receptor: a study with monoclonal antibodies. Biochemistry 22:1811–1819
Whittaker VP (1965) The application of subcellular fractionation techniques to the study of brain function. Prog Biophys Mol Biol 15:39–54

Glutamate and GABA Receptors of Insect Muscles: Biochemical Identification and Interactions with Insecticides

M. E. ELDEFRAWI, I. M. ABALIS, M. T. FILBIN, and A. T. ELDEFRAWI [1]

1 Introduction

Because of the continuous development of insect resistance to insecticides it is important to develop new insecticides which have different structures and different modes of action. The standard approach used by the insecticide industry has been to synthesize chemicals of varying structures and screen them for toxicity to insects. After the proven success of a compound, scientists study its mechanism of action and search for its molecular target. It may be more profitable to reverse the sequence and exploit the wealth of structures of drugs that are known to act on vital proteins in the insect body, as models for synthesis of new generations of insecticides.

Nerve cells use chemical signals (neurotransmitters) to communicate across synapses with each other and with muscular and glandular cells. The neurotransmitter released by one cell is received by another via stereospecific recognition sites on receptor glycoproteins, which are strategically localized mostly in the postsynaptic membranes. Binding of the neurotransmitter to its receptor causes a change in receptor conformation, thereby initiating a change in membrane potential or the production of a second messenger and triggering cellular response. Thus these receptors are vital to nervous function and regulation of muscular activity. The insect nervous system has many of the neurotransmitters and their receptors that are found in the vertebrate nervous system, such as acetylcholine (ACh), dopamine, 5-hydroxytryptamine, γ-aminobutyric acid (GABA) and glutamate. Yet, neurotransmitter receptors, which represent primary targets for many drug actions in mammals, have rarely been exploited as targets for insecticidal action.

There are two major differences in neural regulation of skeletal muscles between vertebrates and invertebrates: One is the cholinergic nature of motor neurons regulating muscle activation in vertebrates but the glutamatergic nature of most excitatory motor nerves in invertebrates (Usherwood 1981). The other is the nature of inhibition of motor neurons in vertebrates through central cholinergic inhibitory interneurons, but the peripheral inhibition of the invertebrate muscle via inhibitory GABA-ergic neurons (Cull-Candy and Miledi 1981, Cull-Candy 1982).

Selective toxicity of a drug that acts on a receptor system may result from differences in receptor localization or pharmacology, as well as in the drug's metab-

1 Department of Pharmacology and Experimental Therapeutics, University of Maryland School of Medicine, Baltimore, Maryland 21201, USA

Approaches to New Leads for Insecticides
Edited by von Keyserlingk et al.

olism. For example, the plant alkaloid *d*-tubocurarine and the α-neurotoxins of elapid snakes (e.g., bungarus and cobra) are extremely poisonous to mammals, causing skeletal muscle paralysis because of inhibition of the nicotinic ACh receptor in these muscles. These agents, however, have no effect on insect muscles which do not contain nicotinic ACh receptors and are nontoxic to insects because of their inability as charged compounds to reach the nicotinic ACh receptors of insect ganglia because of the surrounding permeability barrier (M. E. Eldefrawi and O'Brien 1966, 1967, M. E. Eldefrawi et al. 1968).

Differences have been detected in drug specificities of insect and mammalian receptors. For example, nicotinic ACh receptors in the insect central nervous system have much lower affinities for decamethonium, ACh and other choline esters than in vertebrate muscles (Schmidt-Nielsen et al. 1977; M. E. Eldefrawi and A. T. Eldefrawi 1980, Jones et al. 1981). Also, insect muscarinic ACh receptors have higher affinity for *d*-tubocurarine than mammalian receptors (Haim et al. 1979, Shaker and A. T. Eldefrawi 1981).

Glutamate and GABA receptors of insect muscle are excellent potential targets for selective insecticides because of their peripheral localization, which has also facilitated their electrophysiological investigations. Biochemical identification of neurotransmitter receptors relies on drug-binding studies. Initially, the biochemical approach relies heavily on pharmacological data obtained from physiologic studies; however, once established, binding assays rapidly yield information on drug specificity. Binding assays have become extremely useful in receptor structure-activity relationship studies and are used successfully in the pharmaceutical industry for developing new drugs. The following presentation deals with the biochemical identification of GABA and glutamate receptors of insects, and interactions of insecticides with mammalian GABA receptors as well as the kinds of chemical structures that may have insecticidal action.

2 GABA Receptors

Activation of GABA receptors produces inhibition at a variety of sites within the vertebrate and invertebrate central nervous systems as well as invertebrate neuromuscular junctions. Our concept of a GABA receptor has changed radically during the last 10 years. The simple model of a single recognition site associated with a Cl^- channel has been replaced by a single complex protein having at least three distinct interacting components: one that binds GABA, another that binds benzodiazepine (BZ) tranquilizers, and an ion channel which binds the convulsants picrotoxinin (PTX) and bicyclophosphates and the sedative hypnotic barbiturates (Barb) (Fig. 1). Most of this information is drawn from biochemical studies on mammalian neural tissue (Sigel and Barnard 1984, Olsen et al. 1984). These sites interact allosterically and affect each other's binding through conformational changes. The results of a typical binding experiment on rat brain GABA receptors are shown in Table 1. Specific binding of [^{3}H]-flunitrazepam ([^{3}H]-Flu) to the BZ site (i.e., binding inhibited by high concentrations of diazepam) is potentiated by GABA, and specific binding of [^{3}H]-muscimol to the GABA receptor site is activated by pentobarbital and diazepam but specific binding of [^{35}S]-t-bu-

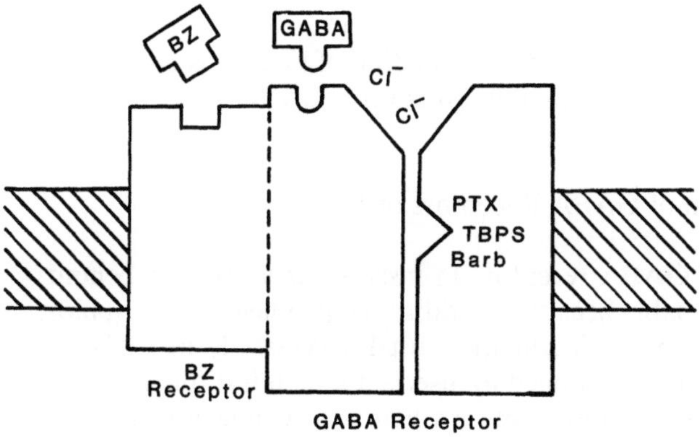

Fig. 1. Schematic representation of a $GABA_A$ type receptor with its Cl^- channel. It has binding sites for GABA and benzodiazepine (*BZ*) and sites that are closely associated with the chloride ion channel: a site that binds picrotoxinin (*PTX*) and *TBPS* and possibly another for the sedative and hypnotic barbiturate (*Barb*)

Table 1. Binding of [^{3}H]-Flu, [^{3}H]-muscimol and [^{35}S]-TBPS to rat brain membranes in presence and absence of specific drug

Radioactive ligand ± drug	Total binding (pmol g^{-1} tissue)	Specific binding (pmol g^{-1} tissue)
[^{3}H]-Flu (1 nM)[a]	6.5±0.2	6.0[b]
[^{3}H]-Flu (1 nM)+3 µM diazepam[a]	0.5±0.01	–
[^{3}H]-Flu (1 nM)+10 µM GABA	12.3±0.3	11.8
[^{3}H]-Muscimol (4 nM)[a]	6.7+0.3	5.6[b]
[^{3}H]-Muscimol (4 nM)+20 µM GABA[a]	1.1±0.06	–
[^{3}H]-Muscimol (4 nM)+100 µM pentobarbital[a]	7.4±0.1	6.3
[^{3}H]-Muscimol (4 nM)+10 µM diazepam[a]	8.4±0.1	7.3
[^{35}S]-TBPS (2 nM)[a]	1.6±0.14	1.3[b]
[^{35}S]-TBPS (2 nM)+5 µM TBPS[a]	0.3±0.09	–
[^{35}S]-TBPS (2 nM)+20 µM PTX[a]	0.4±0.07	–
[^{35}S]-TBPS (2 nM)+10 µM GABA	0.4±0.08	–

[a] Final solution contained 1% ethanol
[b] Specific binding as a % of total binding for [^{3}H]-Flu, [^{3}H]-muscimol and [^{35}S]-TBPS is 92%, 83% and 75% respectively

tylbicyclophosphorothionate ([^{35}S]-TBPS) is inhibited by GABA as previously reported (Squires et al. 1983).

PTX inhibits GABA-induced Cl^- conductances and reverses the GABA enhancement of BZ binding, without inhibiting GABA (Olsen et al. 1978) or BZ (Leeb-Lundberg et al. 1981) binding. In addition, the mammalian neural tissue appears to have endogenous peptides that bind to the GABA and BZ receptor sites (Massotti et al. 1981), but whose role in GABA receptor function is not yet clear.

In mammals there are at least two types of GABA receptors that differ in function, location and drug sensitivity (Enna 1983). Such subclassification of GABA receptors has not been extended to insects. Activation of most GABA receptors leads to an influx of Cl^-, causing hyperpolarization. These "$GABA_A$ receptors" are usually present on neuronal cell body or dendrites, are activated by GABA and 3-amino-l-propanesulfonic acid and are inhibited by bicuculline. $GABA_B$

receptors occur presynaptically on non-GABA ergic neurons, are activated by baclofen, are insensitive to bicuculline and their binding of GABA is dependent upon Ca^{2+} (Bowery et al. 1983). The fluxing ion is probably K^+.

2.1 Biochemical Identification of GABA Receptors of Insects

Biochemical identification of GABA receptors in insects has lagged far behind that of mammalian brain, probably because of additional problems faced in biochemical studies of insect receptors. Techniques used successfully in studying mammalian tissues cannot simply be applied to insects. Often it is difficult to obtain sufficient tissues from insects, and in cases where a mechanical method is used to harvest large amounts of insect heads or thoraces (e.g., houseflies or honeybees) as a source for brains or muscles, the preparation is often contaminated with other tissues. In addition, the high concentration of proteases in insect tissues usually necessitates the use of several antiproteases in the homogenization buffer to protect the receptor protein. Furthermore, a difference in affinity or sparcity of receptor sites sometimes makes it difficult to detect specific binding of radiolabeled probes used successfully with mammalian receptors.

Our initial attempts to measure specific binding of [^{3}H]-GABA or [^{3}H]-muscimol to membranes prepared from housefly heads and thoraces failed because of the high variability and low ratio of specific binding. However, significant amount of specific binding of [^{3}H]-Flu was detectable to thoracic membranes (Fig. 2). Two binding affinities were detected: a K_d of 24.3 nM to 0.2 pmol mg^{-1}

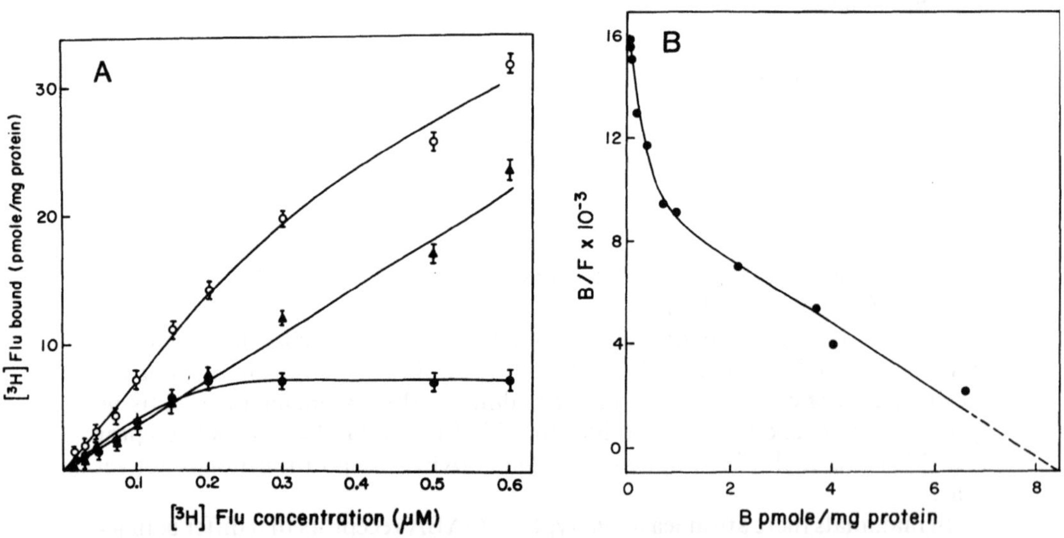

Fig. 2 A, B. Binding of [^{3}H]-Flu to membranes from housefly thoracic muscles. [^{3}H]-Flu at 1 nM was diluted with unlabeled Flu to obtain concentrations up to 1 μM. **A** Saturation isotherm of specific binding (●) calculated as the difference between total binding (○) and non-specific binding (▲) measured in presence of 50 μM diazepam. *Each symbol* and *vertical bar* is the mean ± standard deviation of three experiments. **B** Scatchard plot of the specific binding. *B* represents bound [^{3}H]-Flu in pmoles mg^{-1} protein and *F* is the concentration of free [^{3}H]-Flu in nM. (From Abalis et al. 1983)

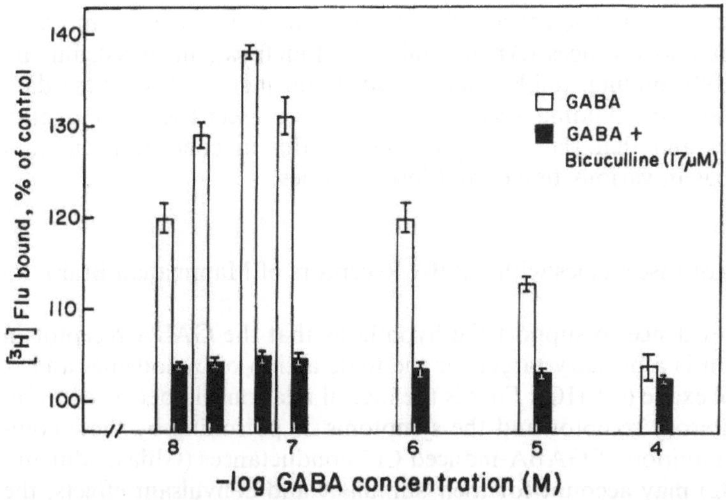

Fig. 3. The stimulation by GABA of 2.5 nM [^{3}H]-Flu binding to membranes from housefly thorax in 5 mM Tris-HCl buffer, pH 7.1, and the inhibition of GABA effect by 17 μM bicuculline. Control 100% level represents the specific binding in absence of *GABA*. *Open columns* represent the specific [^{3}H]-Flu binding in presence of the indicated concentration of *GABA*. *Solid columns* represent binding in presence of both *GABA* and 17 μM bicuculline. *Vertical bars* represent standard deviation of four experiments. (From Abalis et al. 1983)

protein and a K_d of 1 μM to 8.2 pmol mg^{-1} protein (Abalis et al. 1983). These are lower affinities than the K_d of 2.8 nM reported for human brain cortex (Speth et al. 1978) or of 1 nM for rat brain (Braestrup and Squires 1978). Furthermore, we were able to demonstrate coupling of the BZ receptor site to the GABA receptor site by showing a dose-dependent GABA increase of the high affinity [^{3}H]-Flu binding and the inhibition of this increase by 17 μM of the GABA antagonist bicuculline (Fig. 3), a concentration which had no effect on [^{3}H]-Flu binding in absence of other drugs. The high affinity [^{3}H]-Flu binding was also potentiated by other agonists such as muscimol and imidazole acetic acid, but not the GABA uptake inhibitor DL-2,4-diaminobutyric acid (Abalis et al. 1983). Thus, the data suggest the presence of a BZ-binding site on an insect GABA receptor.

In order to characterize this [^{3}H]-Flu binding site, we studied the effect of several BZ's on binding of 2 nM [^{3}H]-Flu to the housefly thoracic membranes. Flu and diazepam were more potent inhibitors than the BZ antagonist ethyl-β-carboline-3-carboxylate or several Roche agonists (Abalis et al. 1983). The very low potency of clonazepam and high potency of Ro-4864 on this binding of [^{3}H]-Flu suggest that the binding specificity is more like that of the mammalian peripheral than the central BZ receptors (Braestrup and Squires 1977, Braestrup et al. 1981). This is not surprising in view of the fact that the housefly membranes that bind [^{3}H]-Flu are most probably from skeletal muscles, since the volume and weight of thoracic muscles were estimated to comprise >98% of thoracic tissues excluding the integument and hemolymph (Filbin et al. 1985), and very little [^{3}H]-Flu-binding was detectable to housefly head membrane preparation mostly from the brain.

On honeybees, we could detect specific [^{3}H]-muscimol binding to membranes from both heads and thoraces (Abalis and A. T. Eldefrawi, in press), but no specific [^{35}S]-TBPS binding, unlike mammalian brain membranes. These data suggest that radioligand binding assays can be used on insect tissues to identify GABA receptors and that these receptors may differ in concentrations and affinities for drugs in various tissues and insect species.

2.2 Interactions of Insecticides with GABA Receptors of Mammalian Brain

There is ample evidence to support the hypothesis that the GABA receptor in mammalian brain is a primary target for the toxic action of cyclodienes and γ-hexachlorocyclohexane (γ-BHC). First is the logical relationship between inhibition of this inhibitory receptor and the symptoms of poisoning by these compounds. Their inhibition of GABA-induced Cl^- conductances (Ghiasuddin and Matsumura 1982) may account for their stimulant and convulsant effects, the same as the convulsants pentamethylenetetrazole and tetrazole analogs, which also displace [^{35}S]-TBPS binding to rat brain GABA receptors (Ramanjaneyulu and Ticku 1984). Second is the very high affinities that the GABA receptor has for these cyclodienes and γ-BHC as shown on [^{3}H]-dihydro-PTX binding (Matsumura and Ghiasuddin 1983) or [^{35}S]-TBPS binding to rat brain membranes (Table 2 and Lawrence and Casida 1984). They inhibit [^{35}S]-TBPS binding in a concentration-dependent manner (Fig. 4) and inhibition by endrin (which at

Table 2. Comparison of the affinity of rat brain GABA receptors for cyclodiene and BHC insecticides to their mammalian toxicity. (From Abalis et al. 1985)

Insecticide	Mammalian toxicity[a] Rat oral LD_{50} ($mg\,kg^{-1}$)	Receptor affinity[b] IC_{50} (nM)
Aldrin	55	500
Dieldrin	46	100
Endrin	10	30
Heptachlor	90	400
Heptachlor epoxide	40	70
Endosulfan I	18	30
Endosulfan II	240	60
γ-BHC	125	150
β-BHC	6,000	> 10,000
Hexachlorobenzene	3,500	> 10,000

[a] Mammalian toxicity values were obtained from EPA-600/2-81-011 Analytical Reference, Standards and Supplemental Data for Pesticides and Other Organic Compounds

[b] Receptor affinity is presented as IC_{50} of binding of 2 nM [^{35}S]-TBPS to rat brain, determined from Figs. 5 and 6. Because the % saturation of binding sites at the concentration is very low, and it appears that binding of these insecticides is to the same site that binds [^{35}S]-TBPS, the K_i is $\simeq$4% lower than the IC_{50}

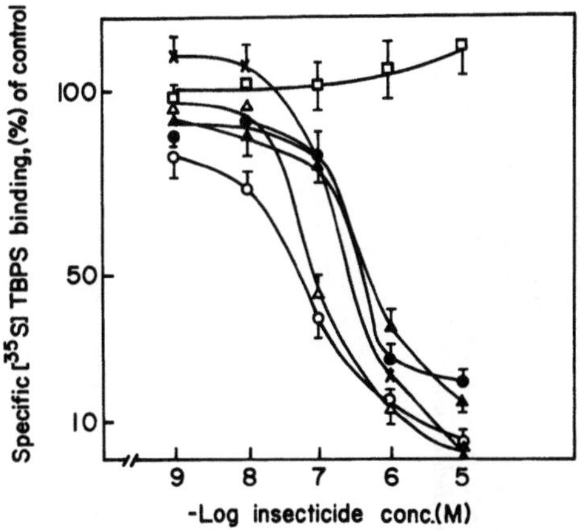

Fig. 4. Inhibition of the specific binding of [^{35}S]-TBPS (2 nM) to GABA receptors in rat brain membranes by PTX (x), two cyclodiene insecticides [heptachlor (●) and aldrin (▲)], their epoxides heptachlor epoxide (○) and dieldrin (△), respectively, and hexachlorobenzene (□). All chemicals were added to the incubation buffer in 10 µl aliquots prior to addition of tissue. Final volume of incubation was 1 ml of KBr-Tris-EDTA-buffer, incubation time was 90 min at 21 °C and 20 mg of original brain tissue were used per assay. Non-specific binding was determined in presence of 20 µM PTX. (From Abalis et al. 1985)

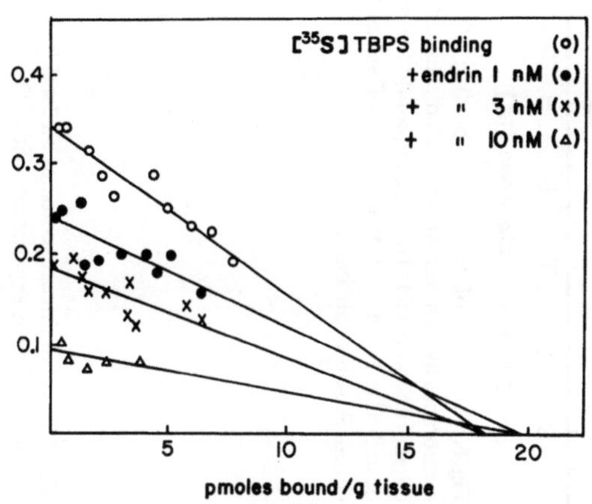

Fig. 5. Scatchard plot of the binding of [^{35}S]-TBPS to the ionic channel site of the GABA receptor of rat brain in absence (○ and presence of 1 nM (●), 3 nM (x) and 10 nM (△) endrin. Data were analyzed by regression analysis. *B* amount bound in pmoles g^{-1} tissue; *F* free [^{35}S]-TBPS concentration. K_d of [^{35}S]-TBPS binding in 51 nM and maximum number of binding sites is 19 ± 2 pmol g^{-1} tissue. (From Abalis et al. 1985). The non-significant difference in the intercept of the lines with the abscissa suggests competitive inhibition and possibly interaction of endrin with the [^{35}S]-TBPS binding site. The K_d of 51 nM is close to those previously reported for [^{35}S]-TBPS binding by Squires et al. (1983) and Lawrence and Casida (1983)

1 nM inhibits 30% of binding) is competitive, suggesting that the cyclodienes and [^{35}S]-TBPS are binding to the same site (Fig. 5). Third is the good correlation for the most part between the potencies of the cyclodienes and BHC isomers in inhibiting [^{35}S]-TBPS binding with their respective mammalian toxicities (Table 2). An exception is in the case of aldrin and lindane, the latter having a lower toxicity but greater affinity than the former. However, these are rat oral LD_{50}'s, and toxicity is a reflection of not only drug-receptor interactions but also drug absorption, distribution, metabolism and elimination. Finally, the best evidence for mammalian brain GABA receptor as a target for cyclodienes and BHC is the stereospecificity of the binding site for diastereoisomers (Fig. 6). Based on the effects of different GABA agonists and antagonists and ion action on [^{35}S]-TBPS binding, it was proposed that this ligand binds to the $GABA_A$, but not the $GABA_B$, receptor complex (Squires et al. 1983).

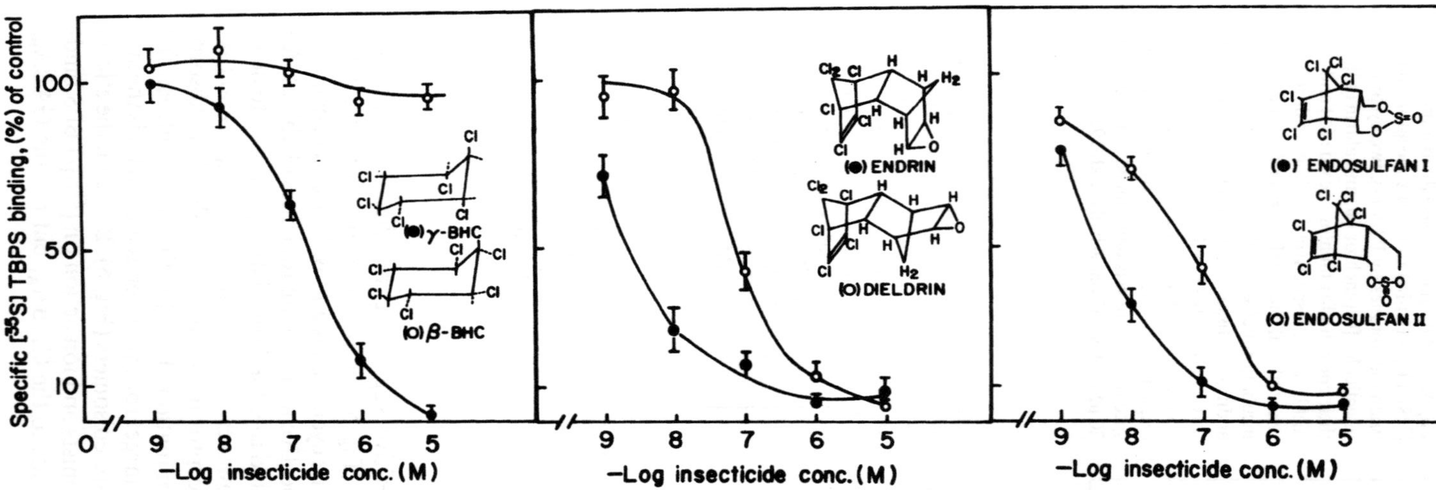

Fig. 6. Inhibition of the specific binding of [^{35}S]-TBPS (2 nM) to GABA receptors in rat brain membranes by stereoisomers of cyclodienes and BHC. All insecticide solutions and [^{35}S]-TBPS were in ethanol and were added in 10 μl aliquots to the incubation tubes. Incubation conditions and time were as described in Fig. 4. Note that in 1 nM of *endosulfan I* and *dieldrin* inhibited 25% and 30%, respectively. (From Abalis et al. 1985)

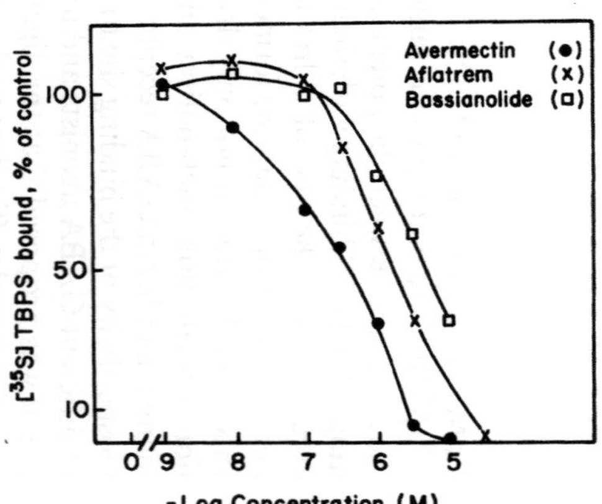

Fig. 7. Effect of three mycotoxins on 2 nM [^{35}S]-TBPS binding to rat brain membranes. Each symbol is the mean of three experiments with SD < 10%

There are several indications that a GABA receptor in insects is also a molecular target for insecticides. BZ's are toxic to aphids though not to the housefly, German cockroach or confused flour beetle (Clifford and Jeffrey 1977), and PTX analogs are potent convulsants on desheathed housefly thoracic ganglion (Miller et al. 1979), though they are non-toxic to houseflies when topically applied (Kuwano et al. 1980). Furthermore, cockroaches resistant to γ-BHC and dieldrin were found to be resistant to PTX (Matsumura and Ghiasuddin 1983), and cyclodienes were potent inhibitors of [^{3}H]-dihydro-Ptx binding to membranes from the nerve cord and brain of the American cockroach (Tanaka et al. 1984). However, despite their high toxicity to mammals (Casida et al. 1976), bicyclophosphates have little toxicity to houseflies (Ozoe et al. 1983), and we have been unable to detect specific [^{35}S]-TBPS binding to housefly or honeybee brain or muscle membranes. Therefore, it appears that the GABA receptors of insects have different drug specificities from those of mammals, although it is still likely that the insecticidal action of cyclodienes is due to their action on an insect GABA receptor.

Other kinds of insecticides have also been proposed to act on GABA receptors such as α-cyano-3-phenoxybenzyl pyrethroids, which inhibit stereospecifically [^{35}S]-TBPS binding to rat brain membranes (Lawrence and Casida 1983). Also avermectin B_{1a}, the anthelmintic drug with insecticidal activity (Ostlind et al. 1979), increases PTX-sensitive chloride conductance and potentiates [3H]-Flu binding (Supavilai and Karobath 1981), thus was proposed to bind to the GABA receptor at a site that might be shared by PTX and Barb (Pong et al. 1982). Indeed we tested the effect of avermectin B_{1a} on [^{35}S]-TBPS binding to rat brain and found it to be inhibitory (Fig. 7) with an IC_{50} of 0.5 µM. However, this inhibition may be allosteric, resulting from binding of avermectin B_{1a} to the GABA binding site, thus acting as a partial agonist (Abalis, M. E. Eldefrawi and A. T. Eldefrawi, unpublished results). In addition, bassianolide, a toxin isolated from an entomopathogenic fungus (Abalis 1981), as well as the tremorgenic aflatrem (Cole 1981), also inhibited [^{35}S]-TBPS binding with IC_{50}'s of 7 µM and 2 µM, respectively.

3 Biochemical Identification of Insect Glutamate Receptors

There are two main types of glutamate receptors in insect skeletal muscle: depolarizing (D) receptors, whose activation increases Na^+ conductance, and hyperpolarizing (H) receptors, which increase Cl^- conductance. The D receptors are ibotenate insensitive and have been found in the central nervous system and in insect muscle, while the H receptors are activated by ibotenate (Usherwood 1980, 1981, Cull-Candy 1982) and were found only extrajunctionally.

Kainic acid was found to allosterically enhance receptor sensitivity to L-glutamate (Nadler 1979), and so did glutamate-containing dipeptides, but they inhibited binding of [^{3}H]-kainic acid to rat brain membranes, leading to the suggestion that there may be a glutamate/kainate/dipeptide receptor complex (Ferkany et al. 1984).

There are a few known competitive glutamate receptor inhibitors, but all have low affinity. There are also other drugs that cause voltage-dependent channel blockade such as *d*-tubocurarine and gallamine (Cull-Candy and Miledi 1983),

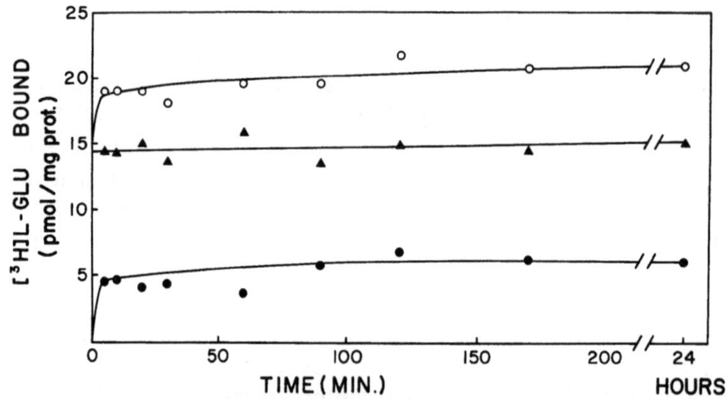

Fig. 8. Binding of [^{3}H]-L-glutamate (100 nM) to housefly thoracic membranes as a function of time. Total binding (○); non-specific binding (▲) and specific binding (●). Results are the means of triplicates with S.E. <8%. (From Filbin et al. 1985)

diltiazem (Ishida and Shinozaki 1980), chlorisondamine (Lingle et al. 1981) and streptomycin (Usherwood 1981) as well as venom components from the *Philanthus* wasp (Clark et al. 1982) and the spiders *Nephila clavata* and *Araneus ventricosus* (Kawai et al. 1982, 1983). These drugs and toxins bind to different sites and allosterically inhibit glutamate receptor function.

We could detect specific [^{3}H]-L-glutamate binding to membranes from insect tissue in thoracic muscles only after the addition of several antiproteases to buffers used in membrane preparation. Binding was rapid (Fig. 8), reversible and saturable (Fig. 9) and sensitive to high heat and proteases (Filbin et al. 1985). It had a high affinity (K_d 0.5 μM), which compares with reported K_d values for mammalian brain putative glutamate receptors of 0.17 μM (Michaelis et al. 1983), 0.3 μM (Sharif and Roberts 1980) and 0.78 μM (Baudry and Lynch 1981). Furthermore, the drug specificity (Table 3) and stereoselectivity (Table 3, Fig. 10) of the binding suggest that the binding molecule is a glutamate receptor.

[^{3}H]-L-Glutamate binding to housefly thoracic muscle membranes was similar to that reported for mammalian brain in that L-glutamate, closely followed by L-aspartate, was the most potent inhibitor of specific binding (Table 3). The putative antagonist, L-glutamate diethylester, was a more potent inhibitor of specific [^{3}H]-L-glutamate binding than the agonist ibotenic acid, which was ineffective on junctional excitatory glutamate receptors in locust muscle (Lea and Usherwood 1973), but was active on those in mammalian brain (Johnston et al. 1974). Also, kainate, which modulates insect glutamate receptors, did not inhibit specific [^{3}H]-L-glutamate-binding to housefly muscle (Table 3) or rat brain (Baudry and Lynch 1979), suggesting that there may be separate kainate receptors (Baudry et al. 1983). Quisqualic acid was a weak inhibitor of [^{3}H]-L-glutamate binding to housefly muscle membranes (Table 3), similar to other reports on rat brain receptor (Biziere et al. 1980, Werling et al. 1983) and different from its agonistic potency on L-glutamate responses in locust muscle (Gration et al. 1981). This discrepancy may be due to a species difference, the kind of muscle, a change in the receptor in situ, or more probably the existence of different receptor types

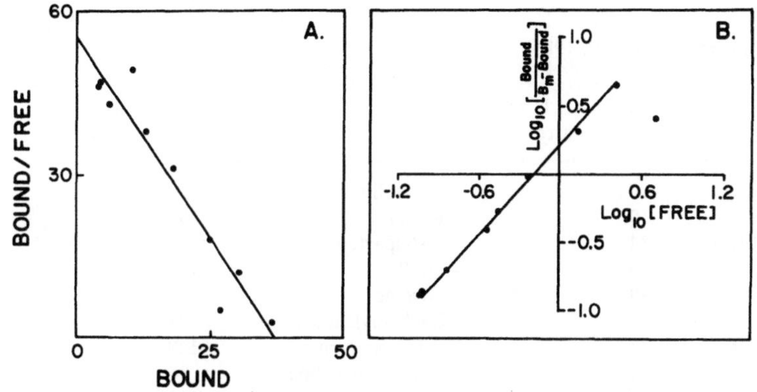

Fig. 9 A, B. Specific [^{3}H]-L-glutamate binding to housefly thoracic membranes. **A** Scatchard analysis of binding data. Bound in pmol mg^{-1} prot. and free in µM. **B** Hill plot of the same data. (From Filbin et al. 1985)

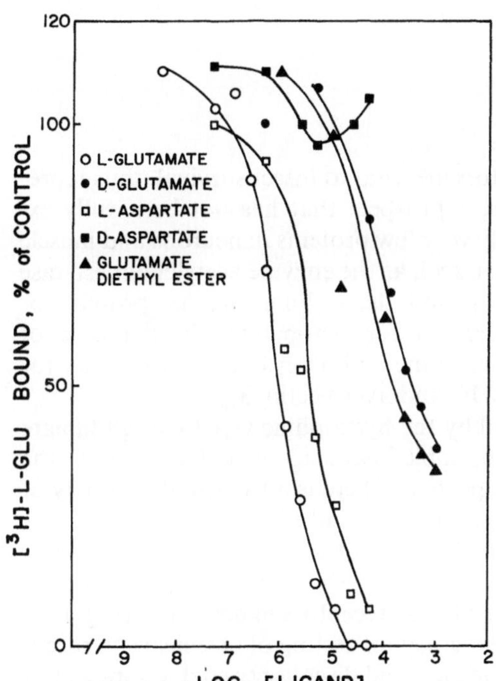

Fig. 10. Inhibition of specific [^{3}H]-L-glutamate (100 nM) binding by various glutamatergic ligands. Incubation was for 2 h at 4 °C. Results are the means of at least two experiments with S.E. <8%. (From Filbin et al. 1985)

which are activated preferentially by glutamate, quisqualate, aspartate or kainate (Butcher et al. 1983, Teichberg et al. 1984). Additional evidence is needed to link this insect [^{3}H]-L-glutamate binding site to glutamate receptor responses, and further improvement of the preparation and modifications of the assay are still needed to develop it into a useful tool for screening drugs and toxins in search of effective blockers of glutamate receptors of insect muscle. Allosteric probes that modulate the properties of the channel of the glutamate receptor would also be most helpful in studies of receptor conformations.

Table 3. Inhibition of specific binding of 100 nM [^{3}H]-L-glutamate to housefly thoracic membranes by various ligands. (From Filbin et al. 1985)

Ligand (1 mM)	[^{3}H]-L-GLU bound % of control	Ligand (1 mM)	[^{3}H]-L-GLU bound % of control
L-Glutamate	0	Proctolin	(92)
L-Aspartate	0	Kainate	(97)
L-Glutamate diethylester	34	D-Aspartate	(100)
D-Glutamate	39	GABA	(105)
N-Methyl-D-aspartate	40	Glycine	84
Ibotenate	52	Carbamylcholine	73
N-Methyl-DL-aspartate	78	*d*-Tubocurarine	128
Quisqualate	84	4-Aminopyridine	86
N-Methyl-L-aspartate	86	Tetraethylammonium	(105)

Results are the means of at least two experiments, each in triplicate (S.E. ± 10%). Values in parenthesis represent non-significant effect

4 Conclusions

Neurotransmitter and hormone receptors are vital to insect survival, thus represent attractive targets for insecticides, a prospect that has not been fully exploited by the insecticide industry. Only very few proteins in neurons and muscle fibers are proven targets for insecticides, such as the enzyme acetylcholinesterase for carbamates and organophosphates, and the sodium channel protein for pyrethroids. There are also the receptors for octopamine in the case of formamidines, for ACh in the case of nicotine and nereistoxin and recently for GABA in the case of cyclodienes, γ-BHC and avermectin B_{1a}.

The fact that a receptor is activated by the hydrophilic GABA or glutamate molecules is no indication that a hydrophobic insecticide would not bind to the same, or different, site on the receptor protein. There is a tremendous variety of structures of chemicals that bind to these receptors and affect their function, particularly because each receptor has a multiplicity of different binding sites. For the GABA receptor alone, the variety is extensive as shown by examples of hydrophobic drugs that are known to bind to GABA receptors in mammals (Fig. 11).

The establishment of a rapid efficient assay for testing of chemicals on receptor binding and function is necessary for the rapid discovery and development of effective insecticides. Receptor-binding assays, which are developed on the basis of electrophysiological findings, are powerful tools for screening of potential insecticides. Already differences are seen between insect and mammalian receptors, as well as between the insect species, which can be utilized to tailor insect-specific poisons.

Acknowledgements. Our reported research was financed in part by NIH grant ES 02594. We are grateful to Dr. R. Cole of the Peanut Institute for providing us with aflatrem, Merck, Sharp & Dohme for avermectin B_{1a}, and the Environmental Protection Agency for cyclodienes and BHC. We thank our secretary Ms. Evelyn Elizabeth for her excellent typing.

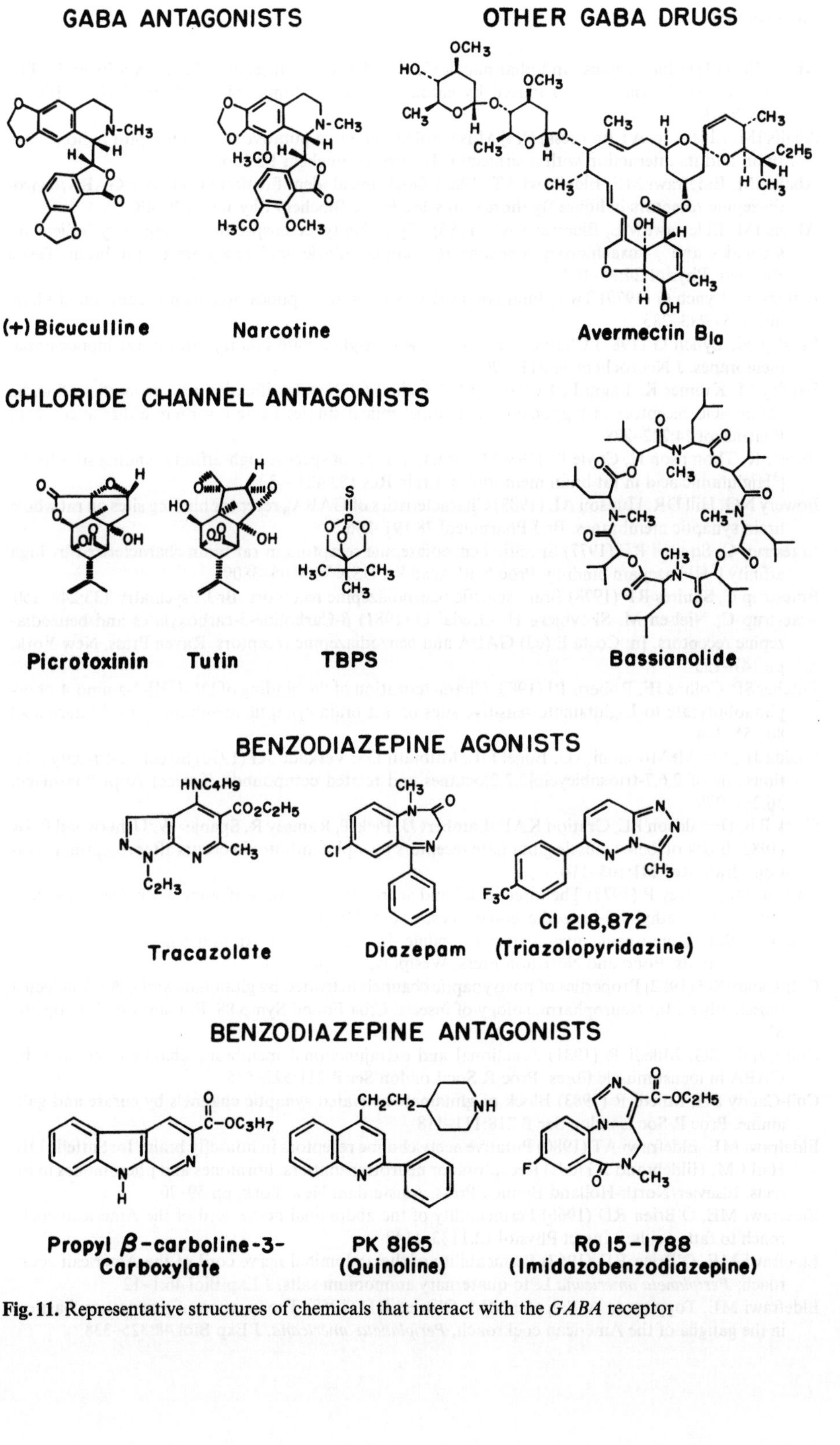

Fig. 11. Representative structures of chemicals that interact with the *GABA* receptor

References

Abalis IM (1981) Biochemical and pharmacological studies of the insecticidal cyclodepsipeptides Destruxins and Bassianolide produced by entomopathogenic fungi. Thesis, Cornell Univ, Ithaca, New York

Abalis IM, Eldefrawi AT (in press) [^{3}H] Muscimol binding to a putative GABA receptor in honey bee brain and its interaction with avermectin B_{1a}. Pestic Biochem Physiol

Abalis IM, Eldefrawi ME, Eldefrawi AT (1983) Biochemical identification of putative GABA/benzodiazepine receptors in house fly thorax muscles. Pestic Biochem Physiol 20:39–48

Abalis IM, Eldefrawi ME, Eldefrawi AT (1985) High affinity stereospecific binding of cyclodiene insecticides and γ-hexachlorocyclohexane to γ-aminobutyric acid receptors of rat brain. Pestic Biochem Physiol 24:95–102

Baudry M, Lynch G (1979) Two glutamate binding sites in rat hippocampal membranes. Eur J Pharmacol 57:283–285

Baudry M, Lynch G (1981) Characterization of two ^{3}H-glutamate binding sites in rat hippocampal membranes. J Neurochem 36:811–820

Baudry M, Kramer K, Fagni L, Recasens M, Lynch G (1983) Classification and properties of acidic amino acid receptors in hippocampus II. Biochemical studies using a sodium efflux assay. Mol Pharmacol 24:222–228

Biziere K, Thompson H, Coyle JT (1980) Characterization of specific high-affinity binding sites for L-[^{3}H]glutamic acid in rat brain membranes. Brain Res 183:421–433

Bowery NG, Hill DR, Hudson AL (1983) Characteristics of $GABA_B$ receptor binding sites on rat whole brain synaptic membranes. Br J Pharmacol 78:191–206

Braestrup C, Squires RF (1977) Specific benzodiazepine receptors in rat brain characterized by high affinity [^{3}H]diazepam binding. Proc Natl Acad Sci USA 74:3805–3809

Braestrup C, Squires RF (1978) Brain specific benzodiazepine receptors. Br J Psychiatry 133:249–260

Braestrup C, Nielsen M, Skovbjerg H, Gredal O (1981) β-Carboline-3-carboxylates and benzodiazepine receptors. In: Costa E (ed) GABA and benzodiazepine receptors. Raven Press, New York, pp 147–155

Butcher SP, Collins JF, Roberts PJ (1983) Characterization of the binding of DL-[^{3}H]-2-amino-4-phosphonobutyrate to L-glutamate-sensitive sites on rat brain synaptic membranes. Br J Pharmacol 80:355–364

Casida JE, Eto M, Moscioni AD, Engel DL, Milbrath DS, Verkade JG (1976) Structure-toxicity relationships of 2,6,7-trioxabicyclo[2.2.2]octanes and related compounds. Toxicol Appl Pharmacol 36:261–279

Clark RB, Donaldson PL, Gration KAF, Lambert JJ, Piek T, Ramsey R, Spanjer W, Usherwood PNR (1982) Block of locust muscle glutamate receptors by δ-philanthotoxin occurs after receptor activation. Brain Res 241:105–114

Clifford DP, Jeffrey P (1977) The insecticidal and acaricidal properties of some 3-alkylcarbamoyloximine-2,4-dimethyl-1,5-benzodiazepines. Pestic Sci 8:446–448

Cole RJ (1981) Tremorgenic mycotoxins: an update. In: Ory RL (ed) Antinutrients and natural toxicants in foods. Food and Nutrition Press, Westport, CT, pp 17–33

Cull-Candy SG (1982) Properties of postsynaptic channels activated by glutamate and GABA in locust muscle fibres. In: Neuropharmacology of insects. Ciba Found Symp 88. Pitman, London, pp 70–82

Cull-Candy SG, Miledi R (1981) Junctional and extrajunctional membrane channels activated by GABA in locust muscle fibres. Proc R Soc London Ser B 211:527–535

Cull-Candy SG, Miledi R (1983) Block of glutamate-activated synaptic channels by curare and gallamine. Proc R Soc London Ser B 218:111–118

Eldefrawi ME, Eldefrawi AT (1980) Putative acetylcholine receptors in housefly brain. In: Sattelle DB, Hall LM, Hildebrand JG (eds) Receptors for neurotransmitters, hormones and pheromones in insects. Elsevier/North-Holland Biomed Press, Amsterdam New York, pp 59–70

Eldefrawi ME, O'Brien RD (1966) Permeability of the abdominal nerve cord of the American cockroach to fatty acids. J Insect Physiol 12:1133–1142

Eldefrawi ME, O'Brien RD (1967) Permeability of the abdominal nerve cord of the American cockroach, *Periplaneta americana* L. to quaternary ammonium salts. J Exp Biol 46:1–12

Eldefrawi ME, Toppozada A, Salpeter MM, O'Brien RD (1968) The location of penetration barriers in the ganglia of the American cockroach, *Periplaneta americana*. J Exp Biol 48:325–338

Enna SJ (1983) Gaba receptors. In: Enna SJ (ed) The GABA receptors. Humana Press, Clifton, NJ, pp 1–23

Ferkany J, Zaczek R, Markl A, Coyle JT (1984) Glutamate-containing dipeptides enhance specific binding at glutamate receptors and inhibit specific binding at kainate receptors in rat brain. Neurosci Lett 44:281–286

Filbin MT, Eldefrawi ME, Eldefrawi AT (1985) Biochemical identification of a putative glutamate receptor in housfly thoracic membranes. Life Sci 36:1531–1539

Ghiasuddin SM, Matsumura F (1982) Inhibition of gamma-aminobutyric acid (GABA)-induced chloride uptake by gamma-BHC and heptachlor epoxide. Comp Biochem Physiol 73 C:141–144

Gration KAF, Lambert JJ, Ramsey R, Usherwood PNR (1981) Nonrandom openings and concentration-dependent lifetimes of glutamate-gated channels in muscle membrane. Nature (London) 291:423–425

Haim N, Nahum S, Dudai Y (1979) Properties of a putative muscarinic cholinergic receptor from *Drosophila melanogaster*. J Neurochem 32:543–552

Ishida M, Shinozaki H (1980) Differential effects of diltiazem on glutamate potentials and excitatory junctional potentials at the crayfish neuromuscular junction. J Physiol (London) 298:301–319

Johnston GAR, Curtis DR, Davis J, McCulloch RM (1974) Spinal interneurone excitation by conformationally restricted analogues of L-glutamic acid. Nature (London) 248:804–805

Jones SW, Sudershan P, O'Brien RD (1981) α-Bungarotoxin binding in house fly heads and *Torpedo* electroplax. J Neurochem 36:447–453

Kawai N, Miwa A, Abe T (1982) Spider venom contains specific blocker of glutaminergic synapses. Brain Res 247:169–171

Kawai N, Miwa A, Abe T (1983) Specific antagonism of the glutamate receptor by an extract from the venom of the spider *Araneus ventricosus*. Toxicon 21:438–440

Kuwano E, Ohshima K, Eto M (1980) Syntheses and insecticidal activity of 8-isopropyl-6-oxabicyclo[3.2.1]octan-7-one, a partial skeleton of picrotoxinin, and related compounds. Agric Biol Chem 44:383–386

Lawrence LJ, Casida JE (1983) Stereospecific action of pyrethroid insecticides on the γ-aminobutyric acid receptor-ionophore complex. Science 221:1399–1401

Lawrence LF, Casida JE (1984) Interactions of lindane, toxaphene and cyclodienes with brain-specific *t*-butylbicyclophosphorothionate receptor. Life Sci 35:171–178

Lea TJ, Usherwood PNR (1973) The site of action of ibotenic acid and the identification of two populations of glutamate receptors on insect muscle-fibres. Comp Gen Pharmacol 4:333–350

Leeb-Lundberg F, Snowman A, Olsen RW (1981) Perturbation of benzodiazepine receptor binding by pyrazolpyridines involves picrotoxinin/barbiturate receptor sites. J Neurosci 1:471–477

Lingle C, Eisen JS, Marder E (1981) Block of glutamatergic excitatory synaptic channels by chlorisondamine. Mol Pharmacol 19:349–353

Massotti M, Guidotti A, Costa E (1981) Characterization of benzodiazepine and γ-aminobutyric acid recognition sites and their endogenous modulators. J Neurosci 1:409–418

Matsumura F, Ghiasuddin SM (1983) Evidence for similarities between cyclodiene type insecticides and picrotoxinin in their action mechanism. J Environ Sci Health B 18:1–14

Michaelis EK, Michaelis ML, Stormann TM, Chittenden WL, Grubbs RD (1983) Purification and molecular characterization of the brain synaptic membrane glutamate-binding protein. J Neurochem 40:1742–1753

Miller TA, Maynard M, Kennedy JM (1979) Structure and insecticidal activity of picrotoxinin analogs. Pestic Biochem Physiol 10:128–136

Nadler JV (1979) Kainic acid: neurophysiological and neurotoxic actions. Life Sci 24:289–300

Olsen RW, Ticku MK, Miller TA (1978) Dihydropicrotoxinin binding to crayfish muscle sites possibly related to γ-aminobutyric acid receptor-ionophore. Mol Pharmacol 14:381–390

Olsen RW, Wong EHF, Stauber GB, King RG (1984) Biochemical pharmacology of the γ-aminobutyric acid receptor/ionophore protein. Fed Proc 43:2773–2778

Ostlind DA, Cifelli S, Lang R (1979) Insecticidal activity of the antiparasitic avermectins. Vet Rec 105:168

Ozoe Y, Mochida K, Nakamura T, Shimizu A, Eto M (1983) Toxicity of bicyclic phosphate GABA antagonists to the housefly *Musca domestica* L. J Pestic Sci 8:601–605

Pong S-S, Dehaven R, Wang CC (1982) A comparative study of avermectin B_{1a} and other modulators of the γ-aminobutyric acid receptor chloride ion channel complex. J Neurosci 2:966–971

Ramanjaneyulu R, Ticku MK (1984) Binding characteristics and interactions of depressant drugs with [^{35}S]*t*-butylbicyclophosphorothionate, a ligand that binds to the picrotoxinin site. J Neurochem 42:221–229

Schmidt-Nielsen BK, Gepner JI, Teng NNH, Hall LM (1977) Characterization of an α-bungarotoxin binding component from *Drosophila melanogaster*. J Neurochem 29:1013–1029

Shaker N, Eldefrawi AT (1981) Muscarinic receptor in housefly brain and its interaction with chlorobenzilate. Pestic Biochem Physiol 15:14–20

Sharif NA, Roberts PJ (1980) Problems associated with the binding of L-glutamic acid to synaptic membranes: methodological aspects. J Neurochem 34:779–784

Sigel E, Barnard EA (1984) A γ-aminobutyric acid/benzodiazepine receptor complex from bovine cerebral cortex. J Biol Chem 259:7219–7223

Speth RC, Wastek GJ, Johnson PC, Yamamura HJ (1978) Benzodiazepine binding in human brain: characterization using [^{3}H]flunitrazepam. Life Sci 22:859–866

Squires RF, Casida JE, Richardson M, Saederup E (1983) [^{35}]*t*-Butylbicyclophosphorothionate binds with high affinity to brain-specific sites coupled to γ-aminobutyric acid-A and ion recognition sites. Mol Pharmacol 23:326–336

Supavilai P, Karobath M (1981) In vitro modulation by avermectin B_{1a} of the GABA/benzodiazepine receptor complex of rat cerebellum. J Neurochem 36:798–803

Tanaka K, Scott JG, Matsumura F (1984) Picrotoxinin receptor in the central nervous system of the American cockroach: its role in the action of cyclodiene-type insecticides. Pestic Biochem Physiol 22:117–127

Teichberg VI, Tal N, Goldberg O, Luini A (1984) Barbiturates, alcohols and the CNS excitatory neurotransmission: specific effects on the kainate and quisqualate receptors. Brain Res 291:285–292

Usherwood PNR (1980) Neuromuscular transmitter receptors of insect muscle. In: Sattelle DB, Hall LH, Hildebrand JG (eds) Receptors for neurotransmitters, hormones and pheromones in insects. Elsevier/North-Holland Biomed Press, Amsterdam New York, pp 141–152

Usherwood PNR (1981) Glutamate synapses and receptors on insect muscle. In: DiChiara G, Gessa GL (eds) Glutamate as a neurotransmitter. Raven Press, New York, pp 183–193

Werling LL, Doman KA, Nadler JV (1983) L-[^{3}H]Glutamate binding to hippocampal synaptic membranes: two binding sites discriminated by their differing affinities for quisqualate. J Neurochem 41:586–593

Biogenic Amines and Second Messenger Systems in Insects

P. D. EVANS[1]

1 Introduction

Biogenic amines have a wide variety of functions in both the central and peripheral nervous systems of insects (Evans 1980). They can act as neurotransmitters, neuromodulators and even circulating neurohormones. Although our knowledge about the pharmacology of the receptors that mediate the actions of biogenic amines in insects is increasing, at present there is only one known example of a pesticide that activates biogenic amine receptors. Our knowledge of the mode of action of insect biogenic amine receptors is also increasing and much evidence has been presented to suggest that many of their actions are mediated through second messenger systems. However, at present no pesticides are known to bring about their actions by directly interfering with second messenger systems in insects.

In the present article I will review the types of biogenic amines found in insect nervous systems and outline the different types of second messenger systems that could be present. I will then discuss in detail two specific examples from insects, namely the action of octopamine receptors on locust muscle and the action of 5-hydroxytryptamine (5-HT) receptors on the blowfly salivary gland, where the second messenger systems present have been described. Finally, I will consider in what ways such systems can offer potential target sites for novel pesticides.

2 Types of Biogenic Amines

A large number of different biogenic amines are present in the insect nervous system (see Evans 1980 for review). The structures of some of the more physiologically important amines are shown in Fig. 1.

Octopamine has a widespread distribution in all insect nervous tissues (Evans 1978, 1985). It is found specifically associated with the cells of the dorsal unpaired median (DUM) cell system (Hoyle 1975, Hoyle and Barker 1975, Evans and O'Shea 1977, 1978). One cell from this group which innervates the extensor-tibiae muscle in the hind leg of the locust acts as a modulatory neurone which can affect neuromuscular transmission (Evans and O'Shea 1977, O'Shea and Evans 1979) and muscular contraction (Evans and Siegler 1982) in this muscle. A neurally evoked release of octopamine from this cell has been demonstrated (Morton and

1 A.F.R.C. Unit of Insect Neurophysiology and Pharmacology, Dept. of Zoology, University of Cambridge, Downing Street, Cambridge CB2 3EJ, United Kingdom

Approaches to New Leads for Insecticides
Edited by von Keyserlingk et al.

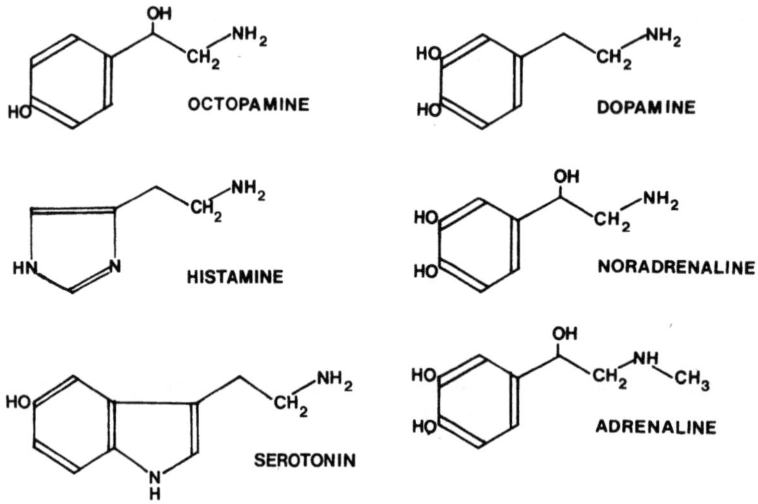

Fig. 1. The structures of some biogenic amines

Evans 1984). Other DUM cells in the abdominal ganglia of fireflies are also octopaminergic (Christensen et al. 1983) and function as excitatory neurones for the light organs (Christensen and Carlson 1981, 1982). In addition octopamine acts as the neurotransmitter of groups of cells in the locust brain that project to the corpora cardiaca where they control the release of a peptide, adipokinetic hormone (Orchard and Loughton 1981, Orchard et al. 1983).

A role as a circulating neurohormone in insect blood has also been suggested for octopamine. The blood levels of octopamine in insects are elevated in the first few minutes of flight (Goosey and Candy 1980) and also under a variety of different stressful conditions (Davenport and Evans 1984a, b). Such elevated levels of haemolymph octopamine would be capable of mobilizing energy reserves from both carbohydrates (Matthews and Downer 1974, Downer 1979a, b) and lipid (Orchard et al. 1981, 1982) stores.

5-hydroxytryptamine (5-HT) is another biogenic amine with a widespread occurrence in insects (see Evans 1980). Recently, the use of immunohistochemical techniques has allowed the identification of a wide range of presumptive serotonergic neurones in the insect central nervous system (e.g. Bishop and O'Shea 1983). Many of these cells may be central interneurones (Taghert and Goodman 1984) although there is evidence that 5-HT may also be contained in some afferent and efferent neurones (Tyrer et al. 1984). Physiologically 5-HT has been shown to be able to induce secretion in the uninnervated salivary glands of the blowfly (Berridge 1972). Currently, it is thought that this action is mediated by increased levels of 5-HT circulating in the haemolymph but the source of this released 5-HT is not known. 5-HT also has modulatory actions on the contractile activity of insect visceral muscles such as those of the heart and gut, but the physiological significance of such effects is not clear at present (see Evans 1980).

Another biogenic amine known to be present in insects is the catecholamine, dopamine. This appears to have a dual role as an intermediary in the tanning processes of insect cuticle and also as a neurotransmitter in the nervous system (Evans 1980). Fluorescent histochemical (Klemm 1976) and radioenzymatic (Dymond and Evans 1979) techniques have shown a widespread distribution of dopamine in the insect nervous system. Recently, these have been complimented by results obtained with high pressure liquid chromatography (Mercer et al. 1983) and immunochemical techniques (Vieillemaringe et al. 1984). There is strong evidence that dopamine acts as a neurotransmitter in cockroach salivary glands (see House 1980). It also has actions on gut and heart muscle, but again the physiological significance of these effects is unclear (Evans 1980).

A second catecholamine present in small quantities in insect nervous tissue is noradrenaline, whilst as yet no convincing evidence has been presented for the presence of adrenaline (Dymond and Evans 1979, Evans 1980). At present the function of noradrenaline in insects is somewhat enigmatic. It is present in only low concentrations in the region of one tenth of the concentration of dopamine present. It is not known if the noradrenaline is present as a metabolic mistake or if a small number of specific noradrenergic neurones exist. It is likely that the development of specific immunochemical techniques for noradrenaline localization would be able to provide an answer to such questions.

Interest has recently focussed on another biogenic amine in the insect nervous system. Histamine is present in insect nervous systems, large concentrations being found particularly in the insect visual system (Elias and Evans 1983). In addition insect nervous tissue can synthesize and metabolize histamine (Elias and Evans 1983) and it is also concentrated against a concentration gradient by a specific set of glial cells in the visual system (Elias and Evans 1984). As yet no functions have been identified for histamine in the insect nervous system.

In studies on the modes of action of the above amines it would be relatively easy if each had its own unique mode of action. However, just as multiple receptor types with different modes of action have been described for various transmitters in the vertebrate nervous system (Snyder and Goodman 1980), it needs to be emphasized that multiple pharmacologically distinct receptor types for biogenic amines also occur in insects. Thus blowfly salivary glands have two distinct types of 5-HT receptor (Berridge and Heslop 1981) and locust muscle has at least three distinct types of octopamine receptor (Evans 1981). It is thus important for each biogenic amine receptor studied in a new insect preparation that the identity of the individual second messenger used in that preparation be determined.

3 Types of Second Messengers

Currently there is much interest in a considerable array of different second messenger systems. However, in insects most of the work in this field has been restricted to a consideration of only two subclasses, namely those that use cyclic AMP and those that use calcium as second messages.

In systems that use cyclic AMP as a second message membrane bound receptors on the outer face of the cell membrane control the interaction of the regulatory and catalytic subunits of adenylate cyclase, the enzyme that synthesizes cyclic AMP (see Rodbell 1984). The increased levels of cyclic AMP resulting from receptor activation cause the activation of specific cyclic AMP dependent protein kinases which bring about the phosphorylation of various target proteins (see Nestler and Greengard 1984). It is the activation of the latter target proteins that bring about the diverse physiological responses of the cells to the applied biogenic amine.

It used to be thought that second messenger systems that used calcium as a message were distinct from those that use other signalling systems such as cyclic GMP, diglyceride and prostaglandins etc. However, recent work suggests that a single unifying signal, namely the breakdown of specific phospholipids in the plasma membrane may be responsible for the activation of a whole complex set of responses (see Berridge 1983, 1984a, b). Further, although I have presented the cyclic AMP and calcium based message systems as separate entities there is a considerable amount of information that suggests they are very intricately linked together (Berridge 1984a, b). Thus many cellular responses to increases in cyclic AMP levels are ultimately expressed as changes in calcium dependent processes, such as changes in calcium permeability, in rates of sequestration or mobilization of calcium or in the sensitivity of other calcium dependent systems (Berridge 1984b).

Since biogenic amine receptor systems are so frequently linked to second messenger systems, it is worth considering what potential advantages this confers on them. There appear to be at least four possible advantages.

First, many of the actions of biogenic amines are of a long time course lasting for several minutes, so that their actions can far outlast the period of their release. Thus in terms of a modulatory action on neuromuscular transmission or the activation of a secretory system it is economical in transmitter terms to produce a long lasting effect using the minimum amount of transmitter, since the exact time of the termination of the response is not usually critical. This does not mean, however, that the response cannot be initiated rapidly.

Second, the use of a second messenger system can lead to the amplification of a weak signal, so that the system can be made sensitive to very small changes in the concentration of the agonist.

Third, the responses can be elicited in regions of the cell at a distance from the site of agonist receptor interactions e.g. on the opposite face of an epithelium, to that which contains the receptors.

Fourth, since there are several steps in the pathway between the binding of the agonist to the receptor and the actual response system, such mechanisms are ideally suited to be modulated by other events, so that the responses may be amplified or inhibited depending perhaps on the arousal state of the insect.

4 Specific Examples from Insects

4.1 Mode of Action of Octopamine Receptors on Locust Muscle

The extensor-tibiae muscle of the hindleg of the locust has many advantages for the study of the mode of action of insect octopamine receptors. It is a "simple" preparation in which all of the four neurones which innervate the muscle have been identified. The extensor muscle is innervated by two excitatory motorneurones, one fast and one slow, together with an inhibitory motorneurone. The muscle is also innervated by an octopaminergic modulatory neurone. The latter neurone is one of a group of dorsal unpaired median neurones and has been christened DUMETi (Hoyle et al. 1974). DUMETi functions to modulate the activity of the motorneurones to this muscle (Evans and O'Shea 1977, O'Shea and Evans 1979, Evans and Siegler 1982) and also inhibits a myogenic rhythm of contractility exhibited by a proximal bundle of muscle fibres (Hoyle 1975, Evans and O'Shea 1978). The effect of DUMETi are long-lasting and far outlast the period of DUMETi stimulation when it is releasing its endogenous octopamine (Morton and Evans 1984).

The effects of octopamine on the extensor muscle are mediated by three pharmacologically distinct classes of octopamine receptor (Evans 1981). OCTOPAMINE$_1$ class receptors mediate the inhibition of the myogenic rhythm whilst

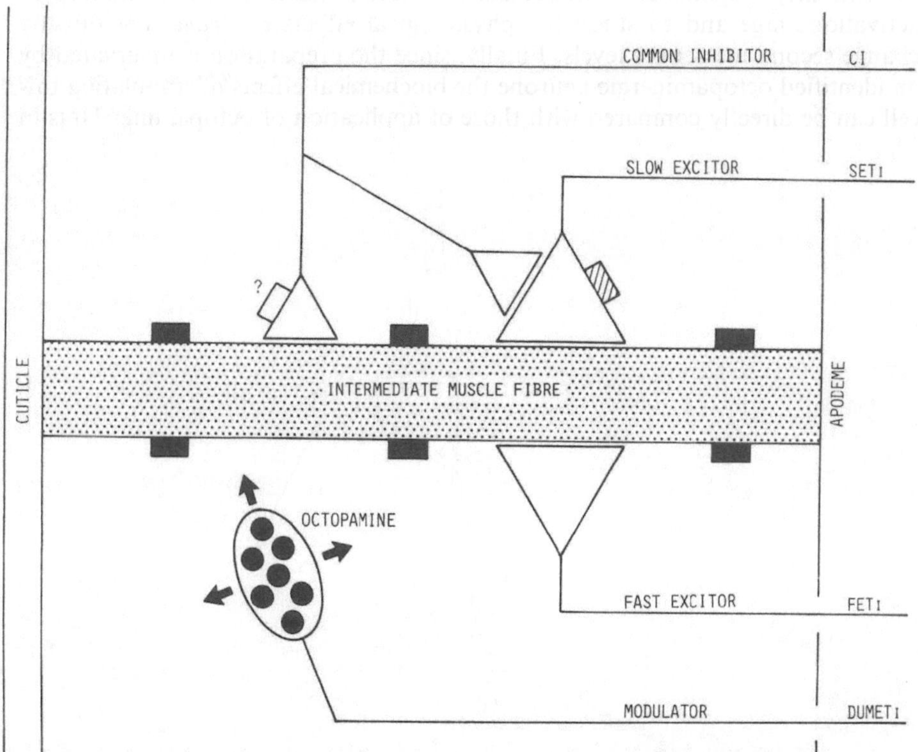

Fig. 2. Diagrammatic representation of the innervation pattern of a muscle fibre from the locust extensor-tibiae muscle to show the distribution of octopamine receptors. OCTOPAMINE$_{2A}$ receptors (*hatched bar*) are located presynaptically on the terminals of the slow motorneurone. OCTOPAMINE$_{2B}$ receptors (*black bars*) are located extrajunctionally on the muscle fibres. There may also be inhibitory octopamine receptors present on the presynaptic terminals of the common inhibitor but these have not yet been characterized. The neurosecretory terminals of the modulatory octopaminergic neurone DUMETi do not form true neuromuscular junctions with the muscle fibres

OCTOPAMINE$_2$ class receptors mediate the modulation of neuromuscular transmission in this muscle. Figure 2 illustrates diagramatically the cellular localization of the latter class of receptors. One subtype of receptor (OCTOPAMINE$_{2A}$) is located presynaptically on the terminals of the slow motorneurone and when stimulated by octopamine, released from the neurosecretory terminals of the DUMETi neurone, potentiates both the spontaneous and neuronally evoked release of neurotransmitter. The second subtype of receptor (OCTOPAMINE$_{2B}$) is located postsynaptically, or extrasynaptically since the DUMETi neurone does not form morphologically distinct neuromuscular junctions, on the muscle fibres themselves. Stimulation of the latter subclass of receptors increases the rate of relaxation of twitch (O'Shea and Evans 1979), tetanic and "catch" tension (Evans and Siegler 1982) in this muscle.

The extensor-tibiae muscle of the locust is also a useful preparation for mode of action studies since it is possible to compare biochemical changes in the muscle with its physiological responses in terms of the time course of the effects, and their susceptibility to agonists and antagonists. It is also possible to bypass the receptor activation stage and to study the physiological effects of drugs that directly change second messenger levels. Finally, since the preparation is innervated by an identified octopaminergic neurone the biochemical effects of stimulating this cell can be directly compared with those of application of octopamine. Thus in

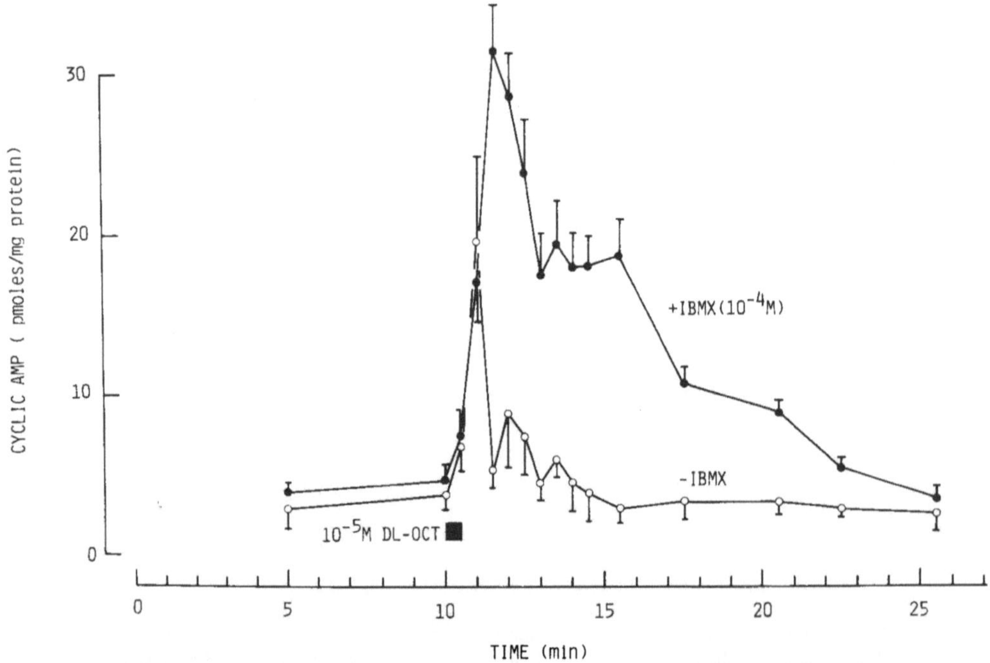

Fig. 3. Time course of 10^{-5} M DL-octopamine mediated increase in cyclic AMP level of extensor-tibiae muscle in the presence (●) and absence (○) of *10^{-4} M IBMX*. The *black bar* indicates time when a 30-s pulse of octopamine was introduced into the muscle superfusate. The results are plotted as pmol cyclic AMP mg^{-1} protein for each of the superfused muscles examined. Each value is the mean of four determinations and the *bars* represent standard errors of the mean. (From Evans 1984a)

this preparation it is possible to meet almost all of the criteria suggested by Beam and Greengard (1976) for the acceptance of a second messenger as the mediator of the effects of activation of a particular class of receptors.

In the extensor muscle the $OCTOPAMINE_2$ class receptors mediate their effects by increasing cyclic AMP levels (Evans 1984a, b). A short pulse of octopamine introduced into the superfusate of the muscle produces a rapid increase in cyclic AMP levels which remain above control values for at least 5 min (Fig. 3). The amplitude of this response and its time course are both potentiated by the phosphodiesterase inhibitor 3 isomethylbutyl xanthine (IBMX). Thus the time course of the biochemical effects are closely matched to that of the physiological responses observed in the muscle.

The increases in cyclic AMP mediated by octopamine in the muscle are dose-dependent (Fig. 4) with a threshold for the effect occurring between 10^{-8} M and 10^{-9} M. By contrast, concentrations of DL-octopamine as high as 10^{-3} M have no significant effect on cyclic GMP levels. The shape of the cyclic AMP dose response curve is somewhat unusual in that the initial rising phase of the sigmoid

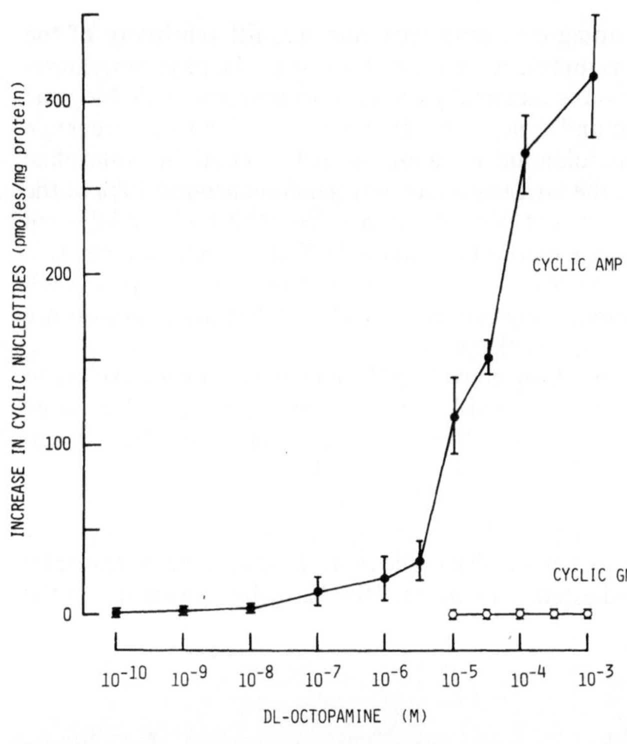

Fig. 4. Dose-response curve for the action of *DL-octopamine* on cyclic nucleotide levels in the extensor-tibiae muscle. The effects on *cyclic AMP* levels (●) and *cyclic GMP* levels (○) are shown. The results are expressed as the increase in cyclic nucleotide levels in pmol mg^{-1} protein in the experimental muscle above that found in the contralateral control muscle. Both experimental and control muscles were pre-incubated for 10 min in 10^{-4} M IBMX before the exposure of the experimental muscle to octopamine plus IBMX for 10 min and the control to a further 10 min incubation in IBMX. Each value is the mean of four determinations and the *bars* represent standard errors of the mean. (From Evans 1984a)

curve extends over more than two log units of concentration before entering the linear portion. In addition the physiological responses of the preparation reach a maximum at a concentration of octopamine around 10^{-5} M, whereas the cyclic AMP response is still rising at this concentration. There are several possible explanations for this observation. One possibility is that octopamine may be acting on more than one receptor type to produce the observed increases in cyclic AMP in the muscle. This could reflect the presence of the $OCTOPAMINE_{2A}$ and $OCTOPAMINE_{2B}$ receptor subtypes in the muscle or even the presence of an additional subtype of receptor that induces an increase in cyclic AMP levels in the muscle but is not related to the observed physiological responses. Another explanation is that by analogy with other receptor systems the extensor muscle may contain "spare receptors" for octopamine (cf. Levitzki 1976, Fain and Berridge 1979). The occupancy of such receptors results in a redundant production of second messengers which is not accompanied by any further increase in the physiological response. It is assumed that spare receptors enable a preparation to respond to lower doses of hormone since the probability of a hormone activating a specific receptor is proportional to the number of receptors on the cell surface (see Berridge 1980).

The amine specificity, antagonist sensitivity and agonist sensitivity of the cyclic AMP responses in this muscle closely match those of the physiological responses. Thus the preparation is maximally sensitive to monophenolic biogenic amines such as octopamine and synephrine. It is also sensitive to α-adrenergic blocking agents such as phentolamine and agonists such as clonidine, tolazoline and naphazoline. However, the latter agonists only generate around 10% of the cyclic AMP made by an equivalent dose of octopamine, whilst in physiological terms they generate responses equal to the maximal effects of octopamine. This suggests that in the extensor muscle it is only necessary to elevate the cyclic AMP levels by 10% to get a maximum physiological effect and that these agonists are only partial agonists at octopamine receptors.

If octopamine receptors work by activating the enzyme adenylate cyclase to increase intracellular levels of cyclic AMP, then increasing cyclic AMP levels by mechanisms that bypass the receptor activation stage should mimic the physiological actions of octopamine. In the extensor muscle elevating intracellular cyclic AMP levels in three different ways mimics all the effects of octopamine receptor activation, including the presynaptic effects of octopamine on the terminals of the slow motorneurone (Evans 1984a, b). The cyclic AMP levels can be increased by application of the phosphodiesterase inhibitor, IBMX, or by application of the

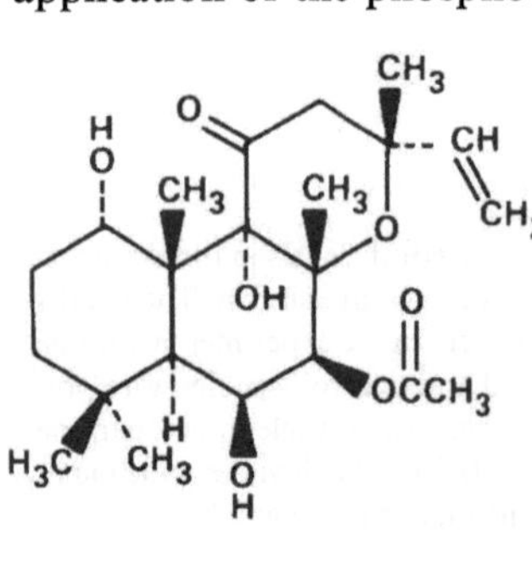

Fig. 5. The structure of forskolin

cyclic AMP analogue 8 chlorophenylthio cyclic AMP, which is more permeable than cyclic AMP itself and more resistant to breakdown by phosphodiesterase. A third method is to expose the preparation to forskolin, a complex diterpene compound (see Fig. 5) which directly activates the catalytic subunit of adenylate cyclase (see Daly 1984). Forskolin specifically increases cyclic AMP levels in the extensor muscle and mimicks all the physiological responses of applying octopamine (Evans 1984a, b).

A diagram summarizing the actions of the $OCTOPAMINE_2$ receptor second messenger systems is shown in Fig. 6. The $OCTOPAMINE_{2A}$ receptor mediated increased in cyclic AMP levels is assumed to increase the calcium permeability of the terminals of the slow motorneurone to increase both the spontaneous and neurally evoked release of transmitter. The $OCTOPAMINE_{2B}$ receptor mediated increase in cyclic AMP levels can be shown to increase the sequestration rate of free calcium into the sarcoplasmic reticulum, which results in an increased rate of relaxation of tension in the muscle.

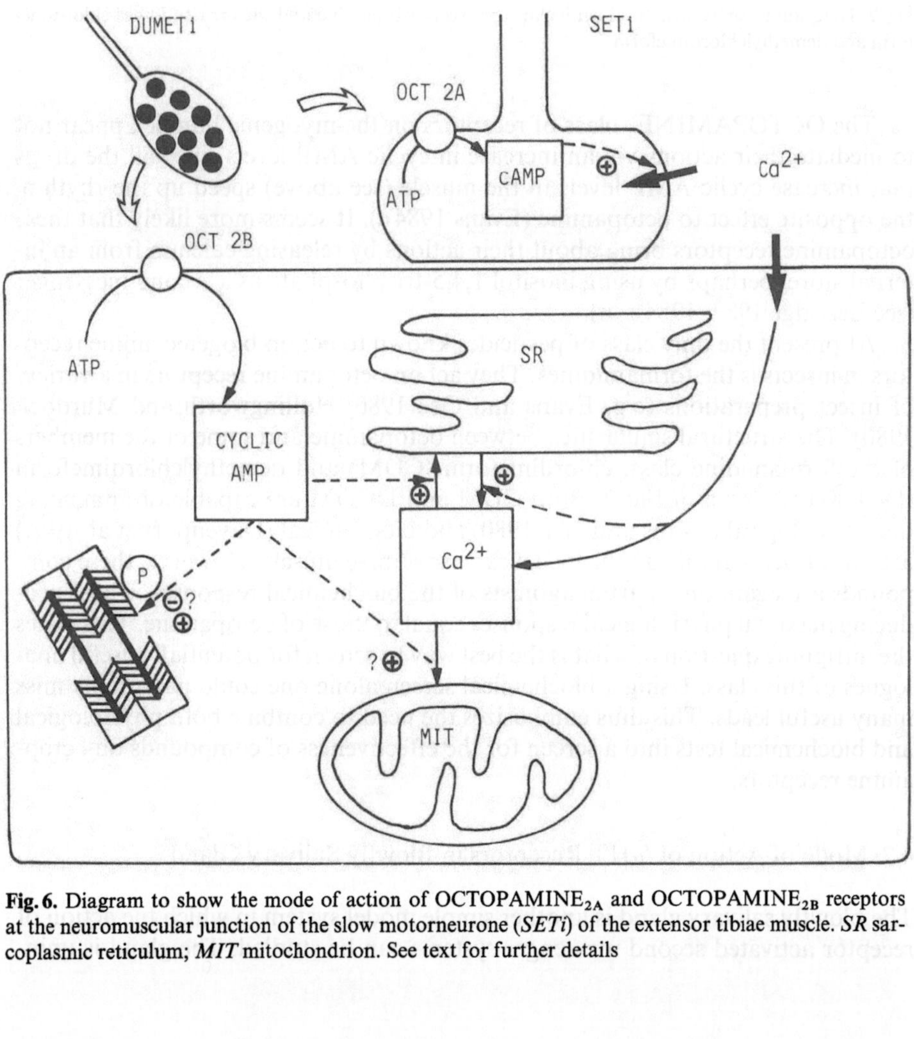

Fig. 6. Diagram to show the mode of action of $OCTOPAMINE_{2A}$ and $OCTOPAMINE_{2B}$ receptors at the neuromuscular junction of the slow motorneurone (*SETi*) of the extensor tibiae muscle. *SR* sarcoplasmic reticulum; *MIT* mitochondrion. See text for further details

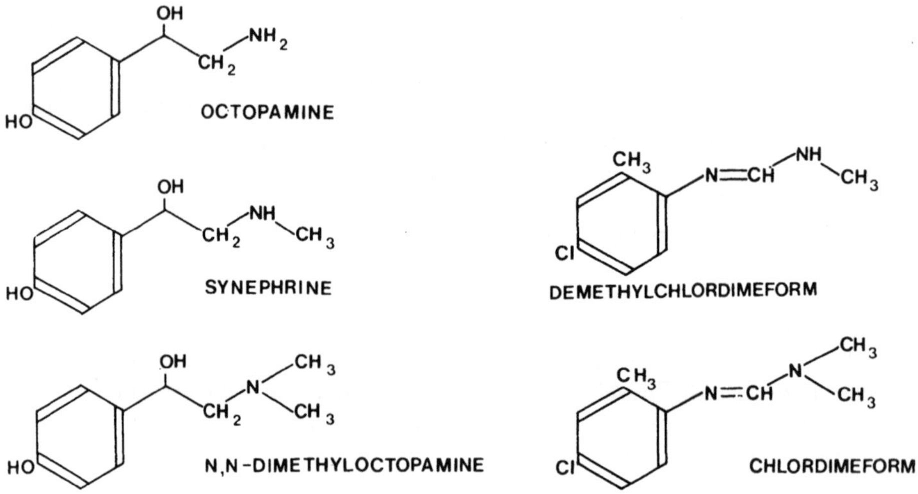

Fig. 7. Diagram to show structural similarities between octopamine and the formamidines chlordimeform and demethylchlordimeform

The OCTOPAMINE$_1$ class of receptors on the myogenic bundle appear not to mediate their actions via an increase in cyclic AMP levels since all the drugs that increase cyclic AMP levels in the muscle (see above) speed up the rhythm, the opposite effect to octopamine (Evans 1984c). It seems more likely that these octopamine receptors bring about their actions by releasing calcium from an internal store, perhaps by using inositol 1,4,5-trisphosphate as a second messenger (see Berridge 1983, 1984a, b).

At present the only class of pesticides known to act on biogenic amine receptors in insects is the formamidines. They act on octopamine receptors in a variety of insect preparations (e.g. Evans and Gee 1980, Hollingworth and Murdock 1980). The structural similarities between octopamine and some of the members of the formamidine class, chlordimeform (CDM) and demethylchlordimeform (DCDM) is shown in Fig. 7. Both CDM and DCDM are capable of mimicking the physiological (Evans and Gee 1980) and biochemical (Davenport et al. 1985) actions of octopamine on the locust extensor tibiae muscle. However, these compounds are again only partial agonists of the biochemical responses, whilst producing maximal physiological responses equal to those of octopamine. This raises the intriguing question of what is the best way to screen for potentially useful analogues of this class. Using a biochemical screen alone one could potentially miss many useful leads. This thus emphasizes the need to combine both physiological and biochemical tests into a screen for the effectiveness of compounds on octopamine receptors.

4.2 Mode of Action of 5-HT Receptors in Blowfly Salivary Gland

The blowfly salivary gland is another simple model system in which the action of receptor activated second messenger systems can be studied. The gland is uninnervated and consists of a tube made from a single layer of epithelial cells in which 5-HT can induce the secretion of isosmotic potassium chloride (Berridge and Prince 1972). Two separate receptor types occur in the gland for 5-HT (Berridge and Heslop 1981). The 5-HT_1 receptor mediates its actions through calcium and the 5-HT_2 receptors mediate its actions through cyclic AMP. It was originally thought that the 5-HT_1 receptor actions depended upon the hydrolysis of phosphatidylinositol in the cell membrane (Berridge and Fain 1979, Fain and Berridge 1979). However, more recent studies (see Berridge 1983, 1984a, b) have shown that the primary action of 5-HT is to stimulate the hydrolysis of phosphatidylinositol 4,5-bisphosphate to yield diacylglycerol and inositol 1,4,5-trisphosphate. The very rapid 5-HT-induced increase in the latter compound suggests that it could function as a second messenger controlling the release of calcium from non-mitochondrial calcium stores (see Fig. 8) (Berridge 1984a, b).

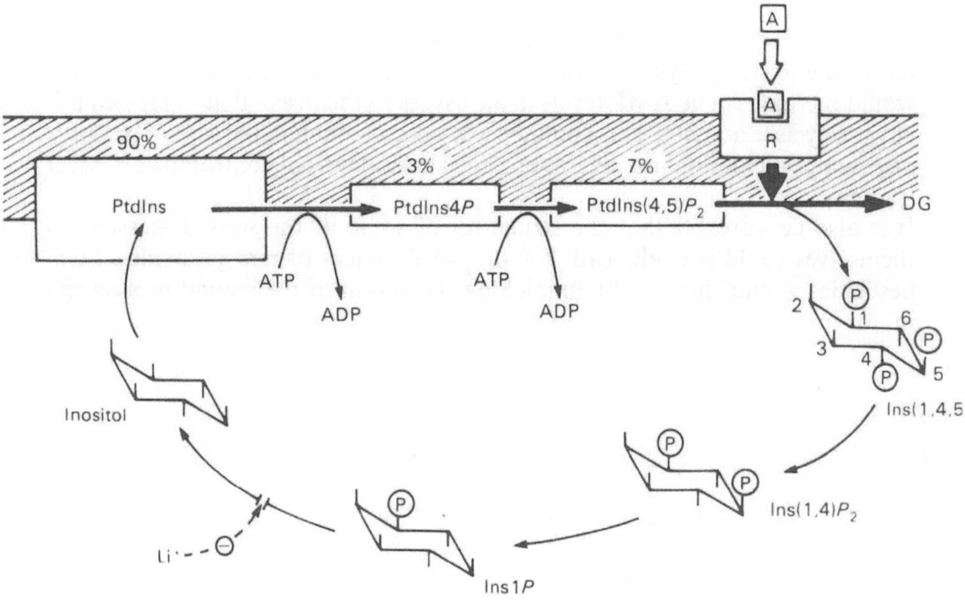

Fig. 8. The mechanism for the agonist (*A*)-dependent hydrolysis of membrane phosphoinositides by Ca^{2+}-mobilizing receptors. The primary substrate for the receptor (*R*) mechanism is the polyphosphoinositide, ***PtdIns (4,5)P₂*** that is hydrolyzed to diglyceride (*DG*) and inositol 1,4,5 trisphosphate [***Ins(1,4,5)P₃***] which then enters an inositol phosphate cycle during which the three phosphates are removed sequentially to liberate the inositol required to resynthesize phosphatidylinositol (***PtdIns***). The final step of this cycle, the dephosphorylation of inositol 1 phosphate (***Ins1P***) to inositol, is inhibited by lithium. The ***PtdIns*** serves as a reservoir to supply precursors that can be phosphorylated to maintain the small hormone sensitive ***PtdIns (4,5)P₂*** pool. (From Berridge 1983)

5 Potential Target Sites for Novel Pesticides

In insects only one class of pesticides, the formamidines, is known to act on biogenic amine receptors and no pesticides are known to directly activate second messenger systems. Thus this area obviously offers a large scope for the development of novel pesticides. Figure 9 illustrates diagramatically some of the potential target sites in such systems where pesticides could act.

1. They could be targeted on to the aminergic receptors themselves. In this case useful lead structures might be obtained from studies on the actions of synthetic agonists or antagonists e.g. the action of substituted imidazolines such as clonidine or tolazoline on octopamine receptors. The advantage of such a site is that it would be selective for those cells in the insect that possess the particular receptor.
2. The various functional subunits of the membrane-bound enzyme systems that generate the second messengers are also potential target sites. For instance, the forskolin-binding site on the catalytic subunit of adenylate cyclase might be a useful site to study. Synthetic pesticides that are developed to activate this site would increase cyclic AMP levels in all tissues in the insect that possessed adenylate cyclase activity. This could be less selective than in the case of receptor oriented pesticides but might still provide a useful way of controlling an insect species.
3. It is also conceivable that the structures of some of the second messengers themselves could provide leads for the development of new pesticides. Such pesticides would then act by mimicking the action of the second messengers in, for instance, the activation of specific protein kinase activities.

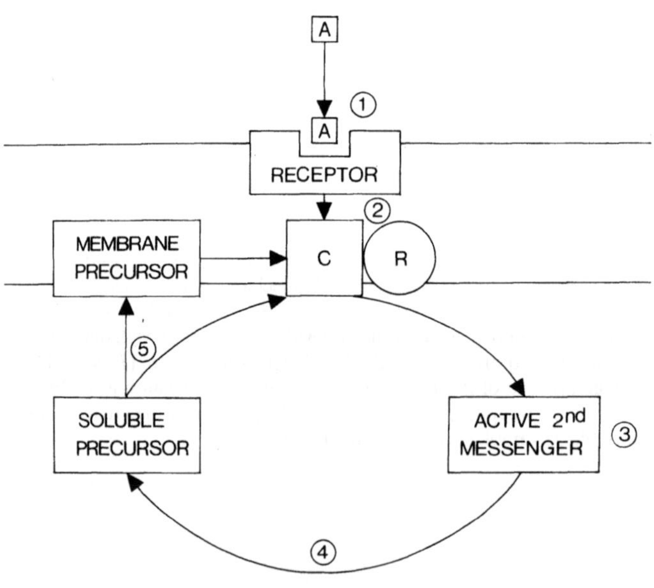

Fig. 9. Potential target sites within a biogenic amine receptor activated second messenger system for the development of novel pesticides. *A* agonist; *C* catalytic subunit; *R* regulatory subunit

4. Another target site in this system could be the enzymes that metabolically inactivate second messengers. Any novel pesticides inhibiting such an activity e.g. the phosphodiesterase enzyme that breaks down cyclic AMP, would again lead to an elevated level of the second messenger.
5. One other potential target site would be the mechanism by which soluble precursors are supplied to the second messenger producing systems in the cell membrane e.g. the supply of inositol for the inositol-containing phospholipids.

Thus biogenic amine activated second messenger systems can provide a wide variety of potential target sites for the development of novel pesticides. At present this is an area in which very little information exists on pesticide action but which will undoubtedly be a very profitable one for future investigation.

References

Beam KG, Greengard P (1976) Cyclic nucleotides, protein phosphorylation and synaptic function. Cold Spring Harbor Symp Quant Biol 40:157–168

Berridge MJ (1972) The mode of action of 5-hydroxytryptamine. J Exp Biol 56:311–321

Berridge MJ (1980) Receptor coupling to adenylate cyclase and calcium gates. In: Sattelle DB (ed) Receptors for neurotransmitters, hormones and pheromones in insects. Elsevier/North-Holland Biomed Press, Amsterdam New York, pp 27–38

Berridge MJ (1983) Rapid accumulation of inositol trisphosphate reveals that agonists hydrolyse polyphosphoinositides instead of phosphatidylinositol. Biochem J 212:849–858

Berridge MJ (1984a) Inositol trisphosphate and diacyglycerol as second messengers. Biochem J 220:345–360

Berridge MJ (1984b) Cellular control through interactions between cyclic nucleotides and calcium. Adv Cyclic Nucleotide Protein Phosphoryl Res 17:329–335

Berridge MJ, Fain JN (1979) Inhibition of phosphatidylinositol synthesis and the inactivation of calcium entry after prolonged exposure of the blowfly salivary gland to 5-hydroxytryptamine. Biochem J 178:59–69

Berridge MJ, Heslop JP (1981) Separate 5-hydroxytryptamine receptors on the salivary gland of the blowfly are linked to the generation of either cyclic adenosine 3',5'-monophosphate or calcium signals. Br J Pharmacol 73:729–738

Berridge MJ, Prince WT (1972) The role of cyclic AMP and calcium in hormone action. Adv Insect Physiol 9:1–49

Bishop CA, O'Shea M (1983) Serotonin immunoreactive neurons in the central nervous system of an insect (*Periplaneta americana*). J Neurobiol 14:251–264

Christensen TA, Carlson AD (1981) Symmetrically organized dorsal unpaired median (DUM) neurones and flash control in the male firefly, *Photuris versicolar*. J Exp Biol 93:133–147

Christensen TA, Carlson AD (1982) The neurophysiology of larval firefly luminescence: direct activation through four bifurcating (DUM) neurons. J Comp Physiol 148:503–514

Christensen TA, Sherman TG, McCaman RE, Carlson AD (1983) Presence of octopamine in firefly photomotor neurons. Neuroscience 9:183–189

Daly JW (1984) Forskolin, adenylate cyclase and cell physiology: An overview. Adv Cyclic Nucleotide Protein Phosphoryl Res 17:81–89

Davenport AP, Evans PD (1984a) Stress-induced changes in the octopamine levels of insect haemolymph. Insect Biochem 14:135–143

Davenport AP, Evans PD (1984b) Changes in haemolymph octopamine levels associated with food deprivation in the locust, *Schistocerca gregaria*. Physiol Entomol 9:269–274

Davenport AP, Morton DB, Evans PD (1985) The action of formamidines on octopamine receptors in the locust. Pestic Biochem Physiol 24:45–52

Downer RGH (1979a) Induction of hypertrehalosemia by excitation in *Periplaneta americana*. J Insect Physiol 25:59–63
Downer RGH (1979b) Trehalose production in isolated fat body of the American cockroach. *Periplaneta americana*. Comp Biochem Physiol 62C:31–34
Dymond GR, Evans PD (1979) Biogenic amines in the nervous system of the cockroach, *Periplaneta americana:* association of octopamine with mushroom bodies and dorsal unpaired median (DUM) neurones. Insect Biochem 9:535–545
Elias MS, Evans PD (1983) Histamine in the insect nervous system: distribution, synthesis and metabolism. J Neurochem 41:562–568
Elias MS, Evans PD (1984) Autoradiographic localization of ^{3}H-histamine accumulation by the visual system of the locust. Cell Tissue Res 238:105–112
Evans PD (1978) Octopamine distribution in the insect nervous system. J Neurochem 30:1009–1013
Evans PD (1980) Biogenic amines in the insect nervous system. Adv Insect Physiol 15:317–473
Evans PD (1981) Multiple receptor types for octopamine in the locust. J Physiol (London) 318:99–122
Evans PD (1984a) Modulation of cyclic nucleotide levels by an octopaminergic neurone in locust skeletal muscle. J Physiol (London) 348:307–324
Evans PD (1984b) The role of cyclic AMP and calcium in the mediation of the modulatory effects of octopamine on locust skeletal muscle. J Physiol (London) 348:325–340
Evans PD (1984c) Studies on the mode of action of octopamine, 5-hydroxytryptamine and proctolin on a myogenic rhythm in the locust. J Exp Biol 110:231–251
Evans PD (1985) Octopamine. In: Kerkut GA, Gilbert LI (eds) Comprehensive insect physiology biochemistry and pharmacology, vol XI. Pergamon Press, Oxford New York, pp 499–530
Evans PD, Gee JD (1980) Action of formamidine pesticides on octopamine receptors. Nature (London) 287:60–62
Evans PD, O'Shea M (1977) An octopaminergic neurone modulates neuromuscular transmission in the locust. Nature (London) 270:257–259
Evans PD, O'Shea M (1978) The identification of an octopaminergic neurone and the modulation of a myogenic rhythm in the locust. J Exp Biol 73:235–260
Evans PD, Siegler MVS (1982) Octopamine mediated relaxation of maintained and catch tension in locust skeletal muscle. J Physiol (London) 324:93–112
Fain JN, Berridge MJ (1979) Relationship between hormonal activation of phosphatidylinositol hydrolysis, fluid secretion and calcium flux in the blowfly salivary gland. Biochem J 178:45–58
Goosey MW, Candy DJ (1980) The D-octopamine content of the haemolymph of the locust *Schistocerca americana gregaria* and its elevation during flight. Insect Biochem 10:393–397
Hollingworth RM, Murdock LL (1980) Formamidine pesticides: octopamine-like actions in a firefly. Science 208:74–76
House CR (1980) Physiology of invertebrate salivary glands. Biol Rev 55:417–473
Hoyle G (1975) Evidence that insect dorsal unpaired median (DUM) neurones are octopaminergic. J Exp Zool 193:425–431
Hoyle G, Barker DL (1975) Synthesis of octopamine by insect dorsal median unpaired neurons. J Exp Zool 193:433–439
Hoyle G, Dagan D, Moberly B, Colquhoun W (1974) Dorsal unpaired median insect neurons make neurosecretory endings on skeletal muscle. J Exp Zool 187:159–165
Klemm N (1976) Histochemistry of putative transmitter substances in the insect brain. Prog Neurobiol 7:99–169
Levitzki A (1976) Catecholamine receptors. In: Cuatrecasas P, Greaves MF (eds) Receptors and recognition, Ser A, vol II. Chapman & Hall, London, pp 199–229
Matthews JR, Downer RGH (1974) Origin of trehalose in stress-induced hyperglycaemia in the american cockroach *Periplaneta americana*. Can J Zool 52:1005–1010
Mercer AR, Mobbs PG, Davenport AP, Evans PD (1983) Biogenic amines in the brain of the honeybee, *Apis mellifera*. Cell Tissue Res 234:655–677
Morton DB, Evans PD (1984) Octopamine realease from an identified neurone in the locust. J Exp Biol 113:269–287
Nestler EJ, Greengard P (1984) Neuron-specific phosphoproteins in mammalian brain. Adv Cyclic Nucleotide Protein Phosphoryl Res 17:483–488

Orchard I, Loughton BG (1981) Is octopamine a transmitter mediating hormone release in insects? J Neurobiol 12:143–153
Orchard I, Loughton BG, Webb RA (1981) Octopamine and short-term hyperlipaemia in the locust. Gen Comp Endocrinol 45:175–180
Orchard I, Carlisle JA, Loughton BG, Gole JWD, Downer RGH (1982) In vitro studies on the effects of octopamine on locust fat body. Gen Comp Endocrinol 48:7–13
Orchard I, Gole JWD, Downer RGH (1983) Pharmacology of aminergic receptors mediating an elevation in cyclic AMP and release of hormone from locust neurosecretory cells. Brain Res 288:349–353
O'Shea M, Evans PD (1979) Potentiation of neuromuscular transmission by an octopaminergic neurone in the locust. J Exp Biol 79:169–190
Rodbell M (1984) Structure-function problems with the adenylate cyclase system. Adv Cyclic Nucleotide Protein Phosphoryl Res 17:207–214
Snyder SH, Goodman RR (1980) Multiple neurotransmitter receptors. J Neurochem 35:5–15
Taghert PH, Goodman CS (1984) Cell determination and differentiation of identified serotonin-immunoreactive neurons in the grasshopper embryo. J Neurosci 4:989–1000
Tyrer NM, Turner JD, Altman JS (1984) Identifiable neurons in the locust central nervous system that react with antibodies to serotonin. J Comp Neurol 227:313–330
Vieillemaringe J, Duris P, Geffard M, Le Moal M, Delaage M, Bensch C, Girardie J (1984) Immunohistochemical localization of dopamine in the brain of the insect *Locusta migratoria migratorioides* in comparison with the catecholamine distribution determined by the histofluorescence technique. Cell Tissue Res 237:391–394

Neuropeptides in Insects: Possible Leads to New Control Methods

M. O'SHEA [1]

1 Introduction

Bioactive peptides are key elements in a complex system of chemical messengers that control all aspects of behavior, metabolism and development in multicellular organisms. It seems clear, therefore, that an improved knowledge of the chemistry and biology of active peptides in insects will aid in the scientific pursuit of safer methods for controlling insects. New methods based on a knowledge of peptide transmitters and hormones, in addition to being safe, may also be relatively insusceptible to the evolution in the pest species of resistance to the insecticide. Considering this potential and the economic importance of insects, it is surprising and unfortunate that of the approximately 20 known invertebrate neuropeptides (O'Shea and Schaffer 1985), only 4 are from insects. These are proctolin (Brown and Starratt 1975), adipokinetic hormone or AKH (Stone et al. 1976) and the two octapeptides MI and MII (Witten et al. 1983, 1984a, O'Shea et al. 1985). Clearly much work remains to be done if the neuropeptide systems are to provide a point of attack for insect control methods. It is particularly important now to obtain more information on the chemical structures of insect peptides. The basic research presents many challenges but the potential for developing new control methods and compounds must be high. There are several reasons for such optimism. First, there are likely to be a large number of different neuropeptides among the insects. Thus the chemist is provided with a new and rich pool of highly bioactive compounds as a base for synthetic approaches to chemical control. Second, a great variety of new, unconventional control methods will become possible with our improved knowledge of active peptides in insects. These methods may involve interference with peptide synthesis, peptide enzymatic inactivation, and peptide release mechanisms, for example. Finally, since peptides are gene products, the enormous potential of applying molecular biological technologies to insect control is opened. Exploiting this option depends on a close collaboration between molecular biologists and insect physiologists and biochemists. Indeed the first steps towards a molecular biological analysis of insect neuropeptides are now being made (Schaffer et al. 1984).

The growth of the molecular approach will have multiple benefits for insect peptide research and pest control. It holds the key for rapid and convenient characterization of new bioactive peptides, for understanding how neuropeptides are

1 Laboratoire de Neurobiologie, Department de Biologie animale, Université de Genève, 1211 Genève 4, Switzerland

Approaches to New Leads for Insecticides
Edited by von Keyserlingk et al.

synthesized and for the analysis of how peptide gene expression is regulated. Along with the scientific advances, there will inevitably be direct benefits to the development of practical applications. In fact it seems unlikely to me that dramatic advances in this applied area will occur in any other way. Precisely how the molecular technologies can be applied, particularly in crop protection, is open to speculation. It may be possible, however, to modify a host plant to produce an important insect neuropeptide hormone or regulator of peptide function in such quantities and in such a form that the development, behavior or metabolism of the target insect feeding on the plant would be compromised.

In this introduction, I have indicated my enthusiasm and hope that new options exist for insect control based on an expanded knowledge of insect peptides. If this is to be more than a hope, much basic scientific work must be done. This will of course be to the benefit of both science and industry and will also require their mutual cooperation. Success offers the hope of improved agricultural production and protection from disease, without the risks associated with the use and manufacture of toxic insecticides. There is no doubt that an urgent need to find safe alternatives to some of the currently used methods exists. The enormity of the tragedy at an insecticide factory in Bhopal, India, unfolding at the time of writing, places a heavy responsibility on us to invest effort and imagination in the development of safe alternatives to the production of highly toxic insecticides. The hormonal and transmitter systems of insects can in my opinion provide some of these new alternatives.

The aims of this brief essay are: (a) to review the current status of research on sequenced neuropeptides in insects; (b) to indicate where more knowledge is needed and how in general it may be obtained and (c) to discuss how this knowledge may eventually be applied.

2 The Sequenced Insect Neuropeptides

There are four neuropeptides in the insects of known amino acid sequence. These are proctolin (Arg-Tyr-Leu-Pro-Thr), adipokinetic hormone or AKH (pGlu-Leu-Asn-Phe-Thr-Pro-Asn-Trp-Gly-Thr-NH_2), and the myoactive peptides MI (pGlu-Val-Asn-Phe-Ser-Pro-Asn-Trp-NH_2) and MII (pGlu-Leu-Thr-Phe-Thr-Pro-Asn-Trp-NH_2).

2.1 Proctolin

Proctolin was first discovered in the American cockroach, *Periplaneta americana* by Brown (Brown 1967). It was shown to have a potent contractile effect on the hindgut or proctodeal muscles and its isolation and purification were guided by the hindgut muscle bioassay (Brown and Starratt 1975). In 1975, Starratt and Brown (Starratt and Brown 1975) succeeded in determining the amino acid sequence of proctolin purified from an extract of about 125,000 cockroaches. Synthetic proctolin is now available from Sigma and Peninsula Labs and a tritiated form is supplied by New England Nuclear.

There exists an extensive literature on the distribution, actions and functions of proctolin. It is arguably the best studied of the invertebrate peptides. It is not, however, the intention of this essay to review this literature in detail but to indicate the general conclusions that can be drawn from studies on proctolin subsequent to its chemical characterization.

The first point is that proctolin is not restricted to the cockroach nor to insects. It is known to occur in at least six orders of insect (Brown 1975), though curiously it is absent from Lepidoptera. More recently, the presence of proctolin has been convincingly demonstrated in crustacea, for example in the lobster *Homarus* (Schwarz et al. 1984) and crayfish (Bishop et al. 1984). Evidence for the existence of proctolin in non-arthropod invertebrates or in vertebrates is at present incomplete. It seems possible, however, that proctolin itself or closely related peptides exist outside the arthropods. In the leech for example, proctolin-immunoreactive neurons have been localized in the central ganglia (Li and Calabrese 1983). In the rat CNS, proctolin immunoreactivity has been reported in serotonin-containing neurons (Holets et al. 1984). It remains to be seen in these examples whether the immunoreactivity is due to the presence of authentic proctolin or proctolin-related peptides.

The second general conclusion from the proctolin literature is that the actions of this peptide are diverse and not restricted to the activity that permitted its isolation. No longer should proctolin be considered to be only a "gut peptide". Once synthetic proctolin became available it was possible to examine its bioactivities in a variety of preparations and this was quickly done. For example, Piek and Mantel (1977) showed that proctolin causes contraction of insect skeletal muscle and induces a myogenic rhythm of contraction and relaxation. Later Miller (1979) demonstrated an action of proctolin on the insect heart. In addition there is now evidence of actions of proctolin on neurons of the insect CNS (Walker et al. 1980). As important as these discoveries were, they were difficult to interpret because the proctolin-sensitive tissues were not demonstrated to be the normal targets of neurally released proctolin. This difficulty was overcome when specific proctolin-containing neurons and their postsynaptic targets were discovered.

Proctolin-containing neurons were found first by application of HPLC methods to neuronal extracts (O'Shea and Adams 1981). Subsequently the availability of synthetic proctolin that permitted physiological studies, also allowed us to develop specific proctolin rabbit antisera (Bishop et al. 1981). This led to the development of a whole-mount immunohistochemical method and its use in the identification of putative proctolinergic neurons in the CNS (Bishop and O'Shea 1982). The application of immunohistochemical methods revealed for the first time the complete complement of proctolin-immunoreactive neurons in the CNS of the cockroach (Bishop and O'Shea 1982) and led to the identification of individual proctolinergic neurons and an analysis of their physiological roles (O'Shea and Bishop 1982, Adams and O'Shea 1983, Keshishian and O'Shea 1985). These studies have revealed the presence in the cockroach, and more recently in the locust (*Schistocerca sp.*), of a subpopulation of proctolinergic skeletal motoneurons. Two such identified neurons, the slow coxal depressor or Ds motoneuron of the cockroach and the slow extensor tibialis or SETi motoneuron of the locust are dual-transmitter neurons (Adams and O'Shea 1983, Witten et al.

1984 b). The peptide functions as a co-transmitter with the usual excitatory transmitter of insect motoneurons, L-glutamic acid. In both examples (Ds and SETi) the released proctolin produces a slow and sustained catch-like muscle contracture when the motoneuron is stimulated. In contrast to the effect of L-glutamic acid, which produces a rapid transient contraction and depolarization, proctolin initiates a long-lasting contracture without depolarizing the muscle. The physiological actions of the proctolinergic motoneurons suggest that an intracellular second messenger mediates the contracture produced by the peptide co-transmitter. These observations establish the third general conclusion, namely that proctolin is involved as a co-transmitter in the process of skeletal neuromuscular transmission, its actions being mediated locally rather than hormonally.

The discovery of peptidergic skeletal motoneurons and the demonstration of a role for a peptide at the insect neuromuscular junction has several implications. For example, it greatly simplifies the problem of investigating neuropeptide function. Electrophysiological and biochemical studies of peptide action which would be impossible in the CNS can conveniently be studied in nerve-muscle preparations. This may be particularly important for the analysis of peptide receptors and the biochemical consequences of peptide action (see Applications, below) because muscle provides a relatively large mass of homogeneous target tissue. We hope that this feature will allow us to characterize the intracellular second messenger or messengers that mediate the actions of proctolin and perhaps other peptides. Our preliminary work on this question in the SETi/extensor tibialis muscle system of the locust has ruled out second messenger roles for cyclic AMP and cyclic GMP. Neither is stimulated at concentrations of proctolin ten times higher than required to produce contracture (Witten and O'Shea, in preparation). Currently our attention is focused on the possibility that proctolin causes the hydrolysis of phosphoinositides which leads to the release of inositol phosphates in the muscle. In other systems, receptors for hormones and transmitters have been shown to liberate inositol 1,4,5-trisphosphate (IP_3) which functions as an intracellular second messenger that mediates the release of calcium from intracellular compartments (see Berridge 1984 for review). Several features of the proctolin-activated contracture suggest a similar role for IP_3 in the locust extensor tibialis muscle and the cockroach coxal depressor muscle.

In the context of developing effective control compounds, it will be significant if insect skeletal muscle proves to be sensitive to a wide variety of insect peptides (most of which will of course remain to be discovered) and may therefore provide convenient preparations for the screening of many potential control compounds. With respect to this, an important question for the future raised by the discovery of proctolinergic skeletal motoneurons is whether peptides other than proctolin are involved in the process of neuromuscular transmission. This possibility seems to me to be rather likely for the following reasons. Proctolin is associated with a rather small subpopulation of skeletal motoneurons (probably less than 5%). If proctolin is the only peptide associated with skeletal neuromuscular transmission, we have certainly been blessed with good fortune because at the time its association with skeletal motoneurons was discovered it happened to be one of only two sequenced insect peptides. It is far more likely that a large number of peptides are involved in insect neuromuscular transmission and that the proctolin-containing subpopulation of motoneurons is merely one of many peptidergic subpopulations. With this idea in mind we have investigated the sensitivity of insect skeletal muscle to other neuropeptides and we are now using skeletal muscle bioassays to screen for likely new peptide neuromuscular transmitters.

This approach has resulted recently in the successful isolation and characterization of two additional insect neuropeptides. These are the myoactive neuropeptides MI and MII, found in the corpus cardiacum (CC) of the cockroach *Periplaneta americana* (Witten et al. 1983, 1984a, O'Shea et al. 1984). These will be discussed in more detail below, but the sensitivity of insect skeletal muscle to these peptides and evidence for their presence in efferent neurons suggest that we can expect to see a greatly expanded number of insect neuromuscular transmitters. Further support for this idea was provided by the recent report of an action for the molluscan peptide FMRFamide in an insect neuromuscular system (Walther et al. 1984). Such developments will of course be important to pure neurobiology but are fundamental also to exploiting possible practical applications of insect neuropeptide research.

2.2 Adipokinetic Hormone

Adipokinetic hormone (AKH) was the name given by Mayer and Candy (Mayer and Candy 1969) to a small peptide located in the CC of the desert locust *Schistocerca gregaria.* The first function assigned to this hormone was the release of diglycerides from the fat body which occurs a few minutes after the initation of flight.

In 1976 the complete chemical characterization of the peptide was reported (Stone et al. 1976). Although the corpus cardiacum represents a highly enriched source of AKH, which permitted a simpler purification scheme than was possible for proctolin, sequencing presented greater problems. This is because the peptide is blocked at both its amino- and carboxy termini, features that presumably protects it from amino- and carboxy peptidase activity in the blood. Complete sequence analysis was achieved by a combination of amino acid analysis, enzymatic digestion, dansyl Edman degradation and low resolution mass spectroscopy (Stone et al. 1976, Stone and Mordue 1980). Synthetic adipokinetic hormone and its $[Tyr]^1$ analogue are currently available from Peninsula Labs.

Recent progress in research on AKH has paralleled in many respects recent advances in our understanding of the neurobiology of proctolin. For example, like proctolin it appears to be a peptide with multiple functions in addition to its well known lipid mobilizing activity. Thus AKH acts on flight muscle to suppress carbohydrate oxidation and stimulate fatty acid oxidation (Robinson and Goldsworthy 1977). It also produces contracture of some locust skeletal muscle at 10^{-8} M (the extensor tibialis) and has an heart acceleratory effect, though at rather high concentration (Stone and Mordue 1980). Further evidence for multiple effects of AKH are provided by its presence in corpora cardiaca of pre-adult instars (it can be detected in 1 st instar or hatchling locusts, unpublished) and by its presence (or AKH-like peptides) in the CNS (Schooneveld et al. 1983). The former suggests functions for AKH associated with behavior other than flight and the latter suggests transmitter as well as hormonal roles. AKH-like peptides are also found in the innervation of insect midgut (unpublished observations) and may therefore, like proctolin, be involved in the control of visceral muscle.

The major impetus to expanded studies on the possible transmitter functions of AKH was the development of AKH antibody by Schooneveld and his associates (Schooneveld et al. 1983). This has provided the first direct evidence for the existence of neurons that may use AKH as a neurotransmitter. Neurons containing AKH-immunoreactivity have not yet been uniquely identified but the fact that AKH can cause skeletal muscle contraction and its presence in the midgut suggests that motoneurons may be among the AKH-ergic neuronal populations. If this is confirmed and if the motoneurons can be uniquely identified we will have a second peptidergic subpopulation of motoneurons. This will support the view that insect motoneurons are functionally and chemically subdivided according to peptide co-transmitter type and will provide an additional stimulus to further physiological studies of peptidergic systems in insects.

2.3 MI and MII Peptides

The cockroach myoactive peptides MI and MII like AKH were isolated from the CC using a bioassay to develop a purification scheme (Witten et al. 1983, O'Shea et al. 1985). The amino acid sequences of these peptides were proposed in 1983 (Witten et al. 1983) based on amino acid analysis, fast atom bombardment mass spectrometry and structure-activity studies. The proposed amino acid sequences were subsequently confirmed by high resolution mass spectrometry and by comparison of natural and synthetic peptides (Witten et al. 1984 a). The structures for MI and MII proposed in 1983 (Witten et al. 1983) have also been confirmed recently by independent studies (Scarborough et al. 1984). These authors unfortunately elected to call these peptides CC1 and CC2 in spite of their demonstrated exact equivalence to the published sequences of the MI and MII peptides. Such arbitrary renaming of known compounds can cause confusion and ought to be avoided as this field expands and gains momentum.

There are striking structural similarities between MI, MII and AKH. In fact, they all appear to belong to a structurally related family which also includes a crustacean peptide called red pigment concentrating hormone or RPCH (Carlsen et al. 1979).

There is little doubt that other members of this family of peptides exist. For example, compound II (Carlsen et al. 1979) from the locust CC and cockroach neurohormone D (Baumann and Gersch 1982) are (evidence from their amino acid compositions) structurally similar to AKH. Actually neurohormone D is almost certainly equivalent to MII. While the phylogenetic and functional significance of peptide families is not understood, their existence can help in the discovery and chemical characterization of new neuropeptides (see below).

We know relatively little about the functional roles played by peptides MI and MII. It is likely, however, that like proctolin and AKH they act both as neurohormones and as transmitters. They are both released from the CC into the hemolymph by a calcium dependent mechanism (O'Shea et al. 1985). As circulating hormones they probably function as hyperglycemic factors (Gade personal communication and 1985). Since they were isolated based on their ability to produce muscle contraction and an acceleration of rhythmic myogenic contractions they are also likely to be involved in the control of skeletal and cardiac muscle. The MI and MII peptides are also present in the CNS (O'Shea et al. 1985).

The development of MI- and MII-specific antibodies will help resolve many of the questions relating to the functions of these peptides. For example, what are the postsynaptic targets of MI/MII-containing neurons? Are they co-localized or are there separate populations of MI- and MII-ergic neurons? Is there a functional difference between the two peptides? At present we cannot distinguish functionally between MI and MII. It seems unlikely, however, that two peptides would be synthesized when one would do, so we can expect to find different specific sites and modes of action for MI and MII.

3 Future Research: Experimental Strategies for New Peptides

In my view an objective of overriding importance now is to expand the library of sequenced insect neuropeptides. This section will be concerned with outlining some experimental strategies that may be helpful in this regard. An excellent and more detailed consideration of the possible approaches to isolation and sequencing, specifically of insect peptides, can be found in Schaffer (1985).

Before considering experimental approaches to the identification of new neuropeptides, it is worth asking why there are currently so few sequenced peptides among the insects. In recent years there has been a very rapid proliferation of newly characterized vertebrate peptides. A parallel expansion in the number of sequenced insect neuropeptides has not occurred and biochemical studies of insect neuropeptides lag far behind the vertebrate field. Part of the reason for this can be attributed to the difficulty of peptide purification and characterization in insects. Compared to vertebrates, the insect source organisms are small and the accumulation of sufficient pure peptide for sequence analysis presents a considerable obstacle. This problem is compounded by the fact that while peptides are likely to be numerous they are rare neuronal phenotypes. Thus any one neuropeptide may be expressed by a very small proportion of the neurons of the nervous system.

The problems presented by the rarity of peptides, however, can and have been overcome. There are presently available techniques for peptide characterization that either circumvent the need for peptide isolation (molecular methods) or conveniently solve the problem of isolating and sequencing rare peptides (HPLC and mass spectrometry, for example). With the major experimental impediments removed, can we now expect a dramatic increase in the number of sequenced neuropeptides from insects? The answer to this question will depend largely on the level of government and industrial funding for insect neuroscience. As important as it surely is, insect neurobiology is currently inadequately funded. This has contributed as much as the purely practical scientific problems to the paucity of sequenced insect peptides compared to vertebrate peptides. Since serious technical impediments no longer exist, there would no doubt be a very rapid increase in the number of sequenced peptides if more money were available for fellowships and for paying for sequencing services and the synthesis of peptides, analogs and oligonucleotides. As a direct consequence of this there would be advances in our understanding of the insect CNS and benefits to insect control research.

The various experimental approaches to peptide identification in insects can be classified into two broad conceptual categories. The first general strategy we can call "activity-directed". In this a biological activity is first detected and an assay based in this activity serves the dual purpose to indicate the presence of a biologically interesting compound and to monitor its purification. In the second strategy, a peptide is identified or isolated based not on a bioactivity but on an idea or hunch that the procedure is likely to select for bioactive peptides. We can call this general strategy "hunch-directed". In its extreme form the peptide can be completely chemically characterized prior to a bioactivity being assigned to it.

3.1 Activity-directed Strategy

The activity-directed strategy is to date by far the most common and is in fact the only one to produce fully characterized neuropeptides in insects. The existing insect neuropeptides therefore provide the best examples of how this strategy works. Proctolin for example was purified from the cockroach (Brown and Starratt 1975) using its physiological activity, contraction of the hindgut to develop and monitor the purification procedures.

An attractive feature of this approach is that prior to purification there is already strong evidence for a functional role of the active compound. Of course, when the activity is first recognized it is not necessarily due to a peptide, but this can be checked by its response to proteases. In order to function well in the isolation of neuropeptides, bioassays ought to be sensitive and specific. These features combine to permit the purification of rare bioactive compounds from complex mixtures represented by a crude extract. It is important to realize that while the bioassay used in the purification of an active peptide might provide a clue or hint to its physiological role, the bioactivity need not be "physiological" and it will almost certainly not represent the only bioactivity of the compound. An example in point is represented by the MI and MII peptides. These were purified from the cockroach (*P. americana*) CC but the bioassay used to detect them initially was developed in the grasshopper (*S. nitens*). The MI and MII peptides are not present in the grasshopper CC and their activity in this species is therefore not strictly "physiological". The assay in the grasshopper was the heart-like pulsatile skeletal muscle associated with the hindleg tibial extensor muscle. The MI and MII peptides cause an acceleration of the beating and a sustained contracture of this muscle, and they are also metabolically active causing an increase in blood lipid. These activities are shared by the MI/MII-like AKH peptide of the grasshopper CC. These cockroach peptides therefore have multiple functions and could have been purified by a different assay, as indeed was AKH.

The purification of MI and MII using a skeletal muscle/heart beat preparation perhaps indicates a useful general guideline for peptide purification. It is well known in vertebrates that cardiac muscle is responsive to a large number of neuroeffector compounds, including peptides. The same may be true in insects. In the insects we have also argued that a variety of peptides are likely to be involved in the control of skeletal muscle. Preparations based on skeletal and/or cardiac muscle may therefore be useful as screening assays in the purification of a number of insect neuropeptides.

Having established a screening bioassay there are a few useful guidelines that can help in the initial preparation of tissue extracts for assay. The first, which was adopted in the purification of AKH, MI and MII and could have simplified the initial purification of proctolin, is to take advantage of the solubility of small peptides in methanol. Significant selection for peptides and rejection of protein can be achieved at the first step of extraction by homogenizing tissue in a high concentration of methanol (we used 90:9:1, methanol, water, acetic acid in the extraction of MI and MII). Another idea, which we have found useful is to take advantage of hydrophobic interactions and pass extracts through a Waters C_{18} reverse phase SepPak prior to bioassay. In one step the SepPak de-salts the extract, removes protein and concentrates a large-volume aqueous extract into a small volume of easily evaporated methanol.

Of course the detection of a biological activity is only the first step in the purification and structural determination of the active compound. That very many activities in the nervous systems of insects exist is well recognized (see Raabe 1983 for review of classical insect neuroendocrinology). Some of these in particular those from *Manduca sexta* (bursicon and eclosion hormone) are chemically well characterized and purified (see Truman and Taghert 1983 for review) and will soon be sequenced. Purification and determination of complete chemical structure may prove very difficult for some, especially for the larger peptides. For the smaller peptides (say <2 Kd) a combination of (reverse phase) HPLC, amino acid analysis and Edman sequencing or FAB mass spectrometry will enable us to assign amino acid sequences. As I indicated at the beginning of this section, whether or not we see these methods widely and routinely applied in entomological studies, with the concomitant proliferation of sequenced insect neuropeptides, depends not on technical developments but on funding levels and cooperation between industry, government and University scientists.

3.2 Hunch-directed Strategy

The second major class of approaches towards the identification of new insect peptides was referred to as hunch-directed. This strategy depends initially on developing methods that, based hopefully on a good idea, are highly likely to indicate new bioactive neuropeptides. The following examples, some hypothetical, fall into this category.

3.2.1 Determine Structure of Precursor Protein

Pursuing the structure of a peptide precursor protein can lead to the discovery of new peptides. Although it may be possible in some cases to isolate and sequence peptide precursors directly, it is more likely that their amino acid sequences will be determined by inference from the nucleotide sequence of the precursor's mRNA. This is because the precursors are synthetic intermediates and likely to be present in vanishingly small amounts and their relatively large size presents a formidable problem for amino acid sequencing. Determining the precursor structure can lead to the discovery of new peptides for two reasons. First, different neuropeptides may occur in a common precursor and second neuropeptides can be recognized in a protein because they are typically flanked by recognizable "processing sequences". Scanning the precursor for processing sequences, two or more basic amino acids for example, or if the peptide is carboxy amidated the sequence gly-x-x (x = lysine or arginine), therefore indicates likely neuropeptide sequences. These can then be synthesized and evidence can be accumulated for their status as neurotransmitters or hormones. For example, antibodies can be raised, bioactivities can be tested, presence in neurons and calcium-dependent release investigated. An example of this approach is provided by the work of Scheller and his associates on the characterization of cDNA (Scheller et al. 1983, Nambu et al. 1983 and Kreiner et al. 1984). The recent revealing of a network of peptidergic neurons in the abdominal ganglion of *Aplysia* using antibodies to amino acid sequences contained in a presumed neuropeptide precursor illustrates its potential usefulness (Kreiner et al. 1984). Another example is provided by the cloning of the *Aplysia* egg laying hormone (ELH) gene (Scheller et al. 1983). The gene sequence strongly suggested the existence of other important peptides. The inferred protein sequence of the ELH precursor contains a number of potential processing sites, suggesting the production of several peptides. One of these, α bag cell peptide (αBCP) has been synthesized and is known to be both bioactive and released from bag cells (Rothman et al. 1983, Sigvardt et al. 1983).

Molecular biological studies of insect neuropeptides are in their infancy. There is little doubt however that studies analogous to those carried out in molluscs can be carried out in insects. In fact work is currently in progress to isolate AKH and proctolin genes (Schaffer et al. 1984). And we hope in the near future to have their nucleotide sequences. There is of course no guarantee that these sequences will encode new neuropeptides but there is certainly sufficient evidence from other studies to suggest this as a high possibility.

Recombinant DNA studies provide the possibility of complete structural characterization of neuropeptide candidates prior to the assigning of any neurobiological function, but they are not the only possibility in the hunch-directed strategy. Among the many other possibilities I have selected the following to offer as examples of the way specific approaches may be devised. They are presented as ideas rather than as specific directions or recipes for success.

3.2.2 Detect Second Messenger Activation

This idea lies somewhere between the activity-directed and the hunch-directed strategy because it depends on a type of bioassay but does not depend on knowing the peptide's biological function. The assumption embodied in the approach is that many biologically active peptides will interact with a receptor acting via a second messenger. If we therefore assay not for a specific biological action but rather for the ability of compounds to elevate levels of second messenger (cAMP, for example), we can exploit a more general characteristic of active peptides. Such an approach, since it depends on a process common to many peptides with different specific bioactivities, could be applied to the isolation and characterization of many biologically active peptides. This imaginative approach was described by Mordue and Morgan (1984) and has been applied in the isolation of diuretic hormones in locusts. It is a particularly attractive idea in this case, and may prove to be so in many others, because isolation of diuretic hormone by more traditional methods (activity-directed) has been hampered by a poor assay of diuretic activity. In this example, Mordue and Morgan measured the elevation of cAMP in Malpighian tubules to detect diuretic hormone. The same assay in different target tissues (muscle, for example) could of course be used to detect and purify novel peptides of as yet unknown biological function. In the future, particularly as new second messengers are discovered and as assays for them are developed, this approach may become more generally important. It may for example be useful not only in the isolation of naturally occurring compounds, but also in screening synthetic ligands.

3.2.3 Select for Released Compounds

The hunch embodied in this strategy is that peptides released from neural or endocrine sources upon depolarization are likely to have biological activity.

An hypothetical experiment may illustrate how this approach could work to simplify the process of isolating new neuropeptides. Suppose, for example, that we wish to search for and isolate new biologically active peptides from the insect gut. Certainly there are good reasons to think that many biologically active peptides exist in the gut's innervation. But the nerves on the gut represent a very small fraction of the gut's mass, which is composed mostly of muscle. If we elect to pursue the conventional activity-directed strategy we must face the fact that a crude extract of gut tissue will be very complex, containing many small peptides of no particular neurobiological interest. The bioactive peptides will represent a very small part of the total mass. The chromatographic profile of such an extract will be extremely complex too. If, however, we were able to chromatograph only those peptides that are released from the gut when the tissue is exposed to saline containing a high concentration of potassium, a selection for biologically active peptides is achieved. The "releasable" cocktail will be far less complex than the total extract but more importantly it will be highly biased towards peptides likely to be of biological interest. The simplified chromatographic profile of the released compounds can then be compared to the complex profile of the total extract to identify the compounds of potential interest.

A technical difficulty with this experiment is that a very small proportion of the total store of peptides are released into a large volume of perfusing saline from which they must be recovered for subsequent chromatography. In our experiments on the potassium-stimulated release of proctolin and the MI and MII peptides for example, we have achieved a maximum release of 10% of the total peptide content. We have devised a simple method for overcoming the problem of recovering peptides present at very low concentrations in the necessarily large volume of the release saline perfusate (see Adams and O'Shea 1983, O'Shea et al. 1985). This involves passing the high potassium release saline through a C_{18} SepPak after it has passed over the experimental tissue. The SepPak functions to trap and thereby concentrate hydrophobic compounds like peptides from the aqueous perfusate. The perfusate can be disgarded, the SepPak can be desalted with distilled water and the peptides eluted into a small volume of methanol (see Fig. 2 of Adams and O'Shea 1983). The methanol can be rapidly evaporated leaving the released and recovered peptides desalted and ready for chromatography or bioassay. By this means peptides present in vanishingly low concentration in large volumes of saline solution can be conveniently recovered and concentrated.

In assessing the potential usefulness of this approach I have considered whether it would have allowed us to identify the MI and MII peptides (which were actually isolated by the conventional assay-directed approach). The answer is yes. An example of a release experiment performed on the CC of *P. americana* is shown in Fig. 2 of O'Shea et al. (1985). It shows clearly that fluorescence and absorbance peaks associated with both MI and MII were detected in a chromatograph (HPLC, C_{18}) of the high potassium saline perfusate but not in the calcium free high potassium perfusate. Thus, had this been the initial experimental strategy, MI and MII would have been identified as being potentially biologically important compounds before a biological activity was assigned to them.

3.2.4 Isolate Peptides with Carboxy Terminal Amides

Another alternative to the traditional bioassay approach has been proposed and successfully used by Tatemoto and Mutt (1978, 1980). They argued that since bioactive peptides are often amidated (as for instance are three of the four known insect hormones), purification specifically based on the amide group could be selective for bioactive peptides. This strategy in which peptides are dansylated prior to chromatographic separation and detection has already led to the detection of two new mammalian neuropeptides, PHI and PYY (Tatemoto and Mutt 1980). According to Schaffer (1985), the same principle could be applied to other peptide modifications such as sulphate derivatives. In considering whether such a strategy could be applied in the characterization of insect peptides, Schaffer points out that the isolation of PHI and PYY were carried out on a very large scale. If the dansyl derivatives of the amidated peptides can be carried out on a complex extract on a picomole scale, then it should certainly be applicable to insect systems. Determination of the function of the peptides identified by this approach will require the isolation of sufficient quantities for sequencing. A synthetic peptide can then be made, antibodies raised and questions of bioactivity and neuronal localization pursued.

3.2.5 Exploit Structural Homology

In this strategy the observation that neuropeptides in different organisms show varying degrees of structural similarity is exploited. In the extreme case, they may be identical. Thus for instance proctolin is found in a wide variety of different insects and crustaceans.

There are several ways to exploit expected homology. For example, the isolation of a closely related but different compounds in different species can be facilitated by using a "non physiological" bioassay. Thus a substance P-like compound was isolated from octopus using a mammalian substance P hypotension assay. This compound, eledoisin, was in fact purified and sequenced before substance P (Erspamer and Anastasi 1962). Unfortunately its function in octopus remains obscure. Several other methods of recognizing new peptides with structural similarity to known peptides can and have been devised. For example, Leung and Stefano (1984) found enkephalins in the mussel (*Mytilus edulis*) using a mammalian receptor binding assay to recognize enkephalin-like compounds in an organism very distantly related to mammals. In addition a wide variety of approaches exploiting immunological recognition can be used. Indeed significant progress has been made using this approach in insect preparations. For example, Duve et al. (1982) have isolated a pancreatic polypeptide-like activity from the blowfly. Certainly there is a wealth of immunological evidence, particularly immunocytochemical, suggesting that insects and other invertebrates contain "vertebrate peptides". Immunological methods will be of help not only in localizing potentially interesting compounds in insects but also in developing purification procedures. The major task now is to pursue these leads to complete structural determination of the vertebrate-like peptides and then to assign a physiological role for the newly characterized peptides.

In this brief section I have tried to indicate in general terms how new insect neuropeptides can be recognized. It should be emphasized, however, that recognition of a new neuropeptide may be of very limited value to the development of new control methods unless it is purified and sequenced.

4 Applications

How can the basic knowledge about insect neuropeptides be applied? This question cannot at present be answered in detail, but some general points can be made and the most likely sites of attack can be identified. We can for instance think of the functioning of a peptidergic system in a way that helps to identify suitable targets for interference with synthetic compounds. In peptidergic systems five functional processes are necessary and these, ordered from the initial synthesis of the peptide to its final functional inactivation after it has been released and performed its physiological role are:

(1) Biosynthesis, (2) Release, (3) Receptor Binding, (4) Induction of Physiological Response, (5) Enzymatic Alteration of Released Peptide. Each of these categories of course represents a complex of events, so within each there may be many possibilities for developing interfering compounds.

In the remaining part of this section I will briefly indicate the possible vulnerable points in peptidergic systems where newly developed control methods may act.

4.1 Peptide Biosynthesis

Neuropeptides represent only part of a gene product. Their complete biosynthesis depends on the enzymatic processing of a precursor protein. Processing occurs presumably in the lumen of the endoplasmic reticulum and in secretory granules. In insects almost nothing is known about the process of neuropeptide biosynthesis: What controls peptide gene expression and what developmental or functional controls operate on the mechanism of transcription, translation, or precursor processing?

All that can be said now is that important possibilities for interference would be opened if such knowledge were obtained. It could lead for example to the development of enzyme inhibitors or compounds that interact with control systems of peptides expression and synthesis.

4.2 Peptide Release

Neuropeptides in insects are secreted both locally as neurotransmitters and into the hemolymph as neurohormones. The release of the peptide neurotransmitter proctolin from the peptidergic Ds motoneuron is controlled by the electrical activity of the neuron. Release itself is calcium dependent and is presumably triggered by a voltage dependent increase in intracellular calcium concentration in the nerve terminals of the Ds neuron (Adams and O'Shea 1983). Peptide release may however also be caused by "releasing factors" which may themselves be peptides, amines or other biologically active compounds. Nutrients circulating in the hemolymph may also act to release metabolically important neuropeptides (see below under Induction of Physiological Response, for further discussion of this). Releasing factors may act presynaptically on the nerve terminals of peptidergic neurons but such processes have not been investigated for peptide neurotransmitters in insects. It is known, however, that the release of peptide neurohormones from the locust corpus cardiacum is controlled by other transmitter substances. For example, Orchard and Loughton (1981) have shown that octopamine can trigger the release into the hemolymph of adipokinetic hormone activity from the locust CC. In the moth *Manduca sexta* the release of eclosion hormone, an as yet unsequenced peptide of about 4.3 Kd, is developmentally regulated by circulating levels of ecdysteroids (Truman 1981). Clearly synthetic compounds with potent effects on the release of peptide neurohormones or transmitters could function to control insect behavior, metabolism and development. But very little is currently known about naturally occurring releasing factors for neuropeptides. A systematic and intensive search for them could be initiated if funds and resources were available.

The experimental problems of detecting released peptides have largely been overcome (see above). It would be a relatively easy matter therefore to design an assay for the ability of both natural and synthetic compounds to release neuropeptides from say the corpus cardiacum. Using such an approach it may be possible to synthesize a systemically or topically active compound that would trigger a massive inappropriate release of an important neurohormone, or alternatively block the action of a natural releasing factor.

4.3 Receptor Binding

After release the peptide binds to a receptor on the cell membrane of the target. It is very unlikely that neuropeptide action will be mediated by non-membrane-bound intracellular receptors. Internalization of bioactive peptides is, however, possible and precedence for internalization does exist. Even in these cases, however, we can expect that the peptide's action will be mediated by binding to an internalized membrane receptor.

We currently know almost nothing about receptors for insect peptides. This is unfortunate because in vertebrate pharmacology the analysis of receptor binding has had a central importance in the development of important synthetic ligands. The development of receptor binding assays for insect neuropeptides will depend on the availability of synthetic peptides radio labelled to a high specific activity. Currently among the sequenced insect peptides only proctolin is supplied as an isotope (^{3}H-Tyr Proctolin). This points to another area where industry could make a significant contribution. Peptide receptor binding assays would of course have great utility in the quest of active compounds that might be useful in insect control. In conjunction with a bioassay this approach could be used to screen peptide analogs for both binding and bioactivity and their ability to act as physiological antagonists.

Much discussion and disagreement surrounds the practical utility in insect control of peptide analogs of neuropeptides. Naturally, except when considering the "molecular option", it is difficult to see how peptides could be useful directly. They are expensive to make and are not able to penetrate insect cuticle. When ingested, unless they are present in enormous concentrations or unless they are protected, they might be too rapidly digested to be useful. Having said this, however, we should not rule out the possibility of developing peptides that would be effective when ingested. This would be especially interesting if a cheap means of delivering them to the feeding insect would also be found. Consider, for example, the existence of a peptide receptor on the lumenal surface of the gut. If an appropriate peptide analog were provided by the host plant in sufficient quantities, the function mediated by that receptor could be attacked.

Nor should we rule out the possibility of developing non-peptide receptor ligands that may be active topically or when ingested. Certainly non-peptide ligands of peptide receptors exist but potentially useful examples, active in insects, will be found only if more natural insect peptides are characterized and receptor binding assays developed. This surely offers one of the best hopes of providing a rational approach to synthesis of potentially useful compounds.

4.4 Induction of Physiological Response

In insects we currently know very little about what happens between the binding of the peptide to its receptor and the induction of a physiological response. The link between receptor binding and physiological response the "transduction mechanism", however, is likely to be associated with intracellular "second messengers". Precedence for this in insects exists for example in the work of Truman and his associates on *Manduca*. This work indicates a second messenger role for cyclic GMP in mediating the behavioral effects of eclosion hormone (Truman et al. 1979). In another example, the action of an unsequenced locust peptide with diuretic activity is thought to act via a receptor that causes elevation of cyclic AMP levels in Malpighian tubules (Mordue and Morgan 1984). It is likely that in the future neuropeptide action in insects will be found to be mediated by second messengers other than cyclic nucleotides. For example, there is growing evidence for a role of inositol trisphosphate and diacylglycerol in mediating actions of hormones (see Berridge 1984 for a review). In this scheme the peptide causes the hydrolysis of a membrane phosphoinositide to liberate diacylglycerol and inositol trisphosphate both of which can function as second messengers. Together these second messengers can initiate a complex cascade of events like mobilization of calcium, activation of protein kinase, release of arachidonic acid and stimulation of cyclic GMP.

It is outside of the scope of this essay to discuss second messenger actions in detail. It seems clear though that an improved understanding of how these transducing systems function in insects can provide possible clues to future control methods. But detailed knowledge in insects concerning which second messengers are involved and how their actions are linked to the physiological response, is unfortunately not available. If compounds that interfer with peptide-induced second messengers or with the multiple events that are subsequently initiated are to be developed, a far more intensive effort in this field must be supported.

As a final speculative note in this section it should be mentioned that access to second messenger systems in insects by the hypothetical control compound may be achieved conveniently by the modification of an ingested nutrient. In support of this idea it was recently reported that a metabolite of glucose induces the closure of potassium channels in the insulin-containing β-cells of the rat pancreas (Ashcroft et al. 1984). Perhaps the glucose metabolite is associated with a cascade of events similar to hormone-activated cascades, with the final physiological event being the modification of the potassium channel protein and the release of insulin. The glucose metabolite might well, for example, be an inositolphosphate. Peptide activated cascades in insects might similarly be accessed and modified by synthetic compounds that could enter the system as nutrients. It would be surprising if systems analogous to that described for the pancreatic β-cells did not also exist in insects.

4.5 Enzymatic Alteration of Released Peptide

After being released, peptides cannot retain the same biological activity indefinitely. Yet again, however, we must say that little is known about the process of neuropeptide inactivation in insects. Borrowing ideas developed by experiments primarily on vertebrates, we can identify enzymatic hydrolysis as an important means of neuropeptide inactivation. In fact some experiments investigating insect peptide metabolism have been performed (Quistad et al. 1984) and this supports the role of an endopeptidase in the biological inactivation of proctolin. While it is more normal to consider enzymatic actions on released peptide as being involved in functional inactivation, it is also possible that functional activation occurs. In this case a peptide is released in a form with relatively low physiological activity and is modified enzymatically to high potency. Precedent for this exists in vertebrates (angiotensin, for example).

The enzymes that alter the physiological potency of released peptides are clearly possible targets for control compounds. But studies on the nature of such enzymes in insects are in their infancy and much more basic research is required if this potentially fruitful approach is to be exploited in developing new insect control compounds.

5 Conclusions

The emphasis of this essay has been to argue for an urgent need greatly to increase the number of sequenced peptides in the insects. I believe that achieving this goal alone will provide the leads and data base necessary for developing a future generation of insect control methods. Many of the ideas expressed here remain merely speculative so long as there are only four sequenced peptides. Any advances in this exciting area must be preceeded by a serious commitment to increasing this number dramatically. I have emphasized that this depends on an increased investment in insect neurobiology and an increased availability to university scientists of the technical and financial resources of the chemical industry. The complete characterization of a new peptide is an expensive undertaking for a university laboratory operating on a typical invertebrate neurobiology grant. It may require, for example, expensive custom synthesis of peptides, radio labelled ligands and oligonucleotides and investment in expensive chromatographic equipment, etc. Biotechnology companies exist now that will undertake amino acid sequencing of pure peptide. But once again these services are out of the financial reach of most of the insect neurobiology laboratories. We are left with an obligatory requirement for the direct involvement of industry in this venture. A relationship between the chemical industry and invertebrate laboratories should develop which is analogous to that which exists now between the pharmaceutical industry and university medical science laboratories. The development of such a commitment would undoubtly bring about an explosion of knowledge which I believe will be the foundation for important future developments in insect control.

Acknowledgements. Supported by funds provided by the University of Geneva. Many of the ideas expressed here were developed during discussions with my colleagues. In particular I wish to thank Martin Schaffer and Jane Witten for allowing me to discuss their ideas and unpublished work and Elisabeth Jeantin for preparing the manuscript.

References

Adams ME, O'Shea M (1983) Peptide cotransmitter at a neuromuscular junction. Science 221:286–289

Ashcroft FM, Harrison DE, Ashcroft SJH (1984) Glucose induces closure of single potassium channels in isolated rat pancreatic α-cells. Nature (London) 312:446–448

Baumann E, Gersch M (1982) Purification and identification of neurohormone D, a heart accelerating peptide from the corpora cardiaca of the cockroach *Periplaneta americana*. Insect Biochem 12:7–14

Berridge MJ (1984) Inositol trisphosphate and diacylglycerol as second messengers. Biochem J 220:345–360

Bishop CA, O'Shea M (1982) Neuropeptide proctolin (H-Arg-Tyr-Leu-Pro-Thr-OH): Immunocytochemical mapping of neurons in the central nervous system of the cockroach. J Comp Neurol 207:223–238

Bishop CA, O'Shea M, Miller RJ (1981) Neuropeptide proctolin (H-Arg-Tyr-Leu-Pro-Thr-OH): Immunological detection and neuronal localization in the insect central nervous system. Proc Natl Acad Sci USA 78:5899–6002

Bishop CA, Wine JJ, O'Shea M (1985) Neuropeptide proctolin in postural motoneurons of the crayfish. J Neurosci 4:2001–2009

Brown BE (1967) Neuromuscular transmitter substance in insect visceral muscle. Science 155:595–597

Brown BE (1975) Proctolin: A peptide transmitter candidate in insects. Life Sci 17:1241–1252

Brown BE, Starratt AN (1975) Isolation of proctolin, a myotropic peptide from *Periplaneta americana*. J Insect Physiol 21:1879–1881

Carlsen J, Christensen M, Josefesson L (1979) Purification and chemical structure of the red pigment-concentrating hormone of the prawn *Leander adspersus*. Gen Comp Endocrinol 30:327–331

Duve H, Thorpe A, Lazarus NR, Lowry PJ (1982) A neuropeptide of the blowfly *Calliphora vomitoria* with an amino acid composition homologous with vertebrate pancreatic polypeptide. Biochem J 201:429–432

Erspamer V, Anastasi A (1962) Structure and pharmacological actions of eledoisin, the active endecapeptide of the posterior salivary gland of Eledone. Experientia 18:58–61

Gade G (1985) Hypertrehalosaemic and myoactive factors from cockroach corpus cardiacum are very likely identical. Naturwissenschaften (in press)

Holets VR, Hokfelt T, Ude J, Eckert M, Hansen S (1984) Coexistence of proctolin with TRH and 5-HT in the rat CNS. Soc Neurosci (Abstr) 10(2):692

Keshishian H, O'Shea M (1985) The distribution of a peptide neuropeptide in the post embryonic grasshopper CNS. J Neurosci 5:992–1004

Kreiner T, Rothbard JB, Schoolnik GK, Scheller RH (1984) Antibodies to synthetic peptides defined by cDNA cloning reveal a network of peptidergic neurons in *Aplysia*. J Neurosci 4(10):2581–2589

Leung MK, Stefano GB (1984) Isolation and identification of enkephalins in pedal ganglia of *Mytilus edulis* (mollusca). Proc Natl Acad Sci USA 81:955–958

Li C, Calabrese RL (1983) Evidence for proctolin-like substances in the central nervous system of the leech. Soc Neurosci (Abstr) 9(1):76

Mayer RJ, Candy DJ (1969) Control of hemolymph lipid concentration during locust flight: An adipokinetic hormone from the corpora cardiaca. J Insect Physiol 15:611–620

Miller T (1979) Nervous vs. neurohormonal control of insect heart beat. Am Zool 19:77–86

Mordue W, Morgan PJ (1984) Diuretic hormones in locusts. Abstr 17 th Int Congr Entomol, Hamburg, p 263

Nambu JR, Taussig R, Mahon AC, Scheller RH (1983) Gene isolation with cDNA probes from identified *Aplysia* neurons: Neuropeptide modulators of cardiovascular physiology. Cell 35:47–56

Orchard I, Loughton BG (1981) Is octopamine a transmitter mediating hormone release in insects? J Neurobiol 12:143–153

O'Shea M, Adams ME (1981) Pentapeptide (Proctolin) associated with an identified neuron. Science 213:567–569

O'Shea M, Bishop CA (1982) Neuropeptide proctolin associated with an identified skeletal motoneuron. J Neurosci 2:1242–1251

O'Shea M, Schaffer MH (1985) Neuropeptide function: The invertebrate contribution. Annu Rev Neurosci 8:171–198

O'Shea M, Witten J, Schaffer M (1985) Isolation and characterization of two myoactive neuropeptides: Further evidence of an invertebrate peptide family. J Neurosci 4:521–529

Piek T, Mantel P (1977) Myogenic contractions in locust muscle induced by proctolin and by wasp, *Philanthus triangulum* venom. J Insect Physiol 23:321–326

Quistad GB, Adams ME, Scarborough RM, Carney RL, Schooley DA (1984) Metabolism of proctolin, a pentapeptide neurotransmitter in insects. Live Sci 34:569–576

Raabe M (1983) Insect neurohormones. Plenum Press, New York London

Robinson NL, Goldsworthy GJ (1977) Adipokinetic hormone and the regulation of carbohydrate and lipid metabolism in a working flight muscle preparation. J Insect Physiol 23:9–16

Rothman BS, Mayeri E, Brown RO, Yuan P-M, Shively JE (1983) Primary structure and neuronal effects of α-bag cell peptide, a second candidate neurotransmitter encoded by a single gene in bag cell neurons of *Aplysia*. Proc Natl Acad Sci USA 80:5753–5757

Scarborough RM, Jamieson GC, Kalish F, Kramer SJ, McEnroe GA, Miller CA, Schooley DA (1984) Isolation and primary structure of two peptides with cardioacceleratory and hyperglycemic activity from the corpora cardiaca of *Periplaneta americana*. Proc Natl Acad Sci USA 81:5575–5579

Schaffer MH (1985) Isolation and characterization of neuropeptides. In: Breer H, Miller TA (eds) Biochemical techniques in insect research. Ser Exp Entomol. Springer, Berlin Heidelberg New York

Schaffer MH, Noyes BE, O'Shea M (1984) Molecular biological studies on the sequenced insect neuropeptides. Soc Neurosci (Abstr) 10(1):152

Scheller RH, Jackson JF, McAllister LB, Rothman BS, Mayeri E, Axel R (1983) A single gene encodes multiple neuropeptides mediating a stereotyped behavior. Cell 32:7–22

Schooneveld H, Tesser GI, Veenstra JA, Romberg-Privee H (1983) Adipokinetic hormone and AKH-like peptide demonstrated in the corpora cardiaca and nervous system of *Locusta migratoria* by immunocytochemistry. Cell Tissue Res 230:67–66

Schwarz TL, Lee GMH, Siwicki KK, Standaert DG, Kravitz EA (1984) Proctolin in the lobster: The distribution, release and characterization of a likely neurohormone. J Neurosci 4:1300–1311

Sigvardt K, Rothman BS, Mayeri E (1983) Analysis of inhibition produced by the candidate neurotransmitter, α-bag cell peptide, in identified neurons of *Aplysia*. Soc Neurosci (Abstr) 9(1):311

Starratt AN, Brown BE (1975) Structure of the pentapeptide proctolin, a proposed neurotransmitter in insects. Life Sci 17:1253–1256

Stone JV, Mordue W (1980) Adipokinetic Hormone. In: Miller TA (ed) Neurohormonal techniques in insects. Springer, Berlin Heidelberg New York, pp 31–80

Stone JV, Mordue W, Batley KE, Morris HR (1976) Structure of locust adipokinetic hormone that regulates lipid utilisation during flight. Nature (London) 263:207–211

Tatemoto K, Mutt V (1978) Chemical determination of polypeptide hormones. Proc Natl Acad Sci USA 75:4115–4119

Tatemoto K, Mutt V (1980) Isolation of two new novel candidate hormones using a chemical method for finding naturally occurring polypeptides. Nature (London) 285:417–418

Truman JW (1981) Interaction between ecdysteroid, eclosion hormone, and bursicon titers in *Manduca sexta*. Am Yool 21:655–661

Truman JW, Taghert PH (1983) Neuropeptides in insects. In: Krieger DT, Brownstein MJ, Martin JB (eds) Brain peptides. Wiley, New York, pp 165–182

Truman JW, Mumby SM, Welch SK (1979) Involvement of cyclic GMP in the release of stereotyped behaviour patterns in moths by a peptide hormone. J Exp Biol 84:201–212

Walker RJ, James VA, Roberts CJ, Kerkut GA (1980) Neurotransmitter receptors in invertebrates. In: Sattelle DB, Hall LM, Hildebrand JG (eds) Receptors for neurotransmitters, hormones and pheromones in insects. Elsevier/North-Holland Biomed Press, Amsterdam New York, pp 41–57

Walther C, Schiebe M, Voigt KH (1984) Synaptic and nonsynaptic effects of molluscan cardioexcitatory neuropetides on locust skeletal muscle. Neurosci Lett 45:99–104

Witten J, Schaffer MH, O'Shea M (1983) Structure and biology of two new related neuropeptides from insect: Further evidence for a peptide family. Soc Neurosci (Abstr) 9(1):313

Witten J, Schaffer MH, O'Shea M, Carter Cook J, Hemling ME, Rinehart KL Jr (1984a) Structures of two cockroach neuropeptides assigned by fast atom bombardment mass spectrometry. Biochem Biophys Res Commun 124:350–358

Witten J, Worden MK, Schaffer MH, O'Shea M (1984b) New classification of insect motoneurons: Expression of different peptide transmitters. Soc Neurosci (Abstr) 10(1):151

Subject Index

MIX
Papier aus verantwortungsvollen Quellen
Paper from responsible sources
FSC® C105338

If you have any concerns about our products, you can contact us on
ProductSafety@springernature.com

In case Publisher is established outside the EU, the EU authorized representative is:
Springer Nature Customer Service Center GmbH
Europaplatz 3, 69115 Heidelberg, Germany

Printed by Libri Plureos GmbH
in Hamburg, Germany